Table of Contents

Graphing Calculator Manual

for

Precalculus: Concepts in Context

Second Edition

Marsha Davis
Eastern Connecticut State University

Judith Flagg Moran
Trinity College

Mary Murphy
Smith College

THOMSON

BROOKS/COLE

Australia • Canada • Mexico • Singapore • Spain • United Kingdom • United States

Printed in the United States of America
1 2 3 4 5 6 7 07 06 05 04 03

Printer: Patterson Printing Company

ISBN: 0-534-37822-6

For more information about our products,
contact us at:
**Thomson Learning Academic Resource Center
1-800-423-0563**

**For permission to use material from this text,
contact us by:**
Phone: 1-800-730-2214
Fax: 1-800-731-2215
Web: http://www.thomsonrights.com

Brooks/Cole—Thomson Learning
10 Davis Drive
Belmont, CA 94002-3098
USA

Asia
Thomson Learning
5 Shenton Way #01-01
UIC Building
Singapore 068808

Australia/New Zealand
Thomson Learning
102 Dodds Street
Southbank, Victoria 3006
Australia

Canada
Nelson
1120 Birchmount Road
Toronto, Ontario M1K 5G4
Canada

Europe/Middle East/South Africa
Thomson Learning
High Holborn House
50/51 Bedford Row
London WC1R 4LR
United Kingdom

Latin America
Thomson Learning
Seneca, 53
Colonia Polanco
11560 Mexico D.F.
Mexico

Spain/Portugal
Paraninfo
Calle/Magallanes, 25
28015 Madrid, Spain

Preface

Graphs and tables play an important role in providing understanding of the mathematical concepts contained in *Precalculus: Concepts in Context, Second Edition*. With the aid of a graphing calculator, students can graph functions, even fairly complex functions, quickly and easily. They can evaluate functions at specific input values and make tables of function values. All of the labs, most of the projects, and many of the exercises in this textbook require the use of graphing technology. The *Graphing Calculator Manual* provides keystroke-level calculator commands and instructions that are useful in working through the labs, projects, and exercises of *Precalculus: Concepts in Context*.

The *Manual* is three books in one: the first provides instructions for the **TI-83Plus** graphing calculator, the second, for the **TI-85/86**, and the third, for the **TI-89/92Plus/Voyage 200**. Each guide begins with a basic tutorial followed by eight chapters corresponding to chapters in *Precalculus: Concepts in Context*. In most cases, the same examples are used for all three calculator guides. The major exception to this parallel structure occurs in the guide for the TI-89/92Plus/Voyage 200 calculator. This guide includes support for this calculator's computer algebra system.

After a basic tutorial, the *Manual* introduces new calculator features chapter by chapter as they are needed. We suggest that students work through the basic tutorial either in class or as a homework assignment. We do not assume that students or instructors will work through the entire *Manual* chronologically. Therefore, we've structured the *Manual* so that after completing the basic tutorial, you can jump around in the *Manual* from topic to topic. Students should work through sections in the *Manual* that pertain to the chapter they are working on. Thus, new calculator features are learned at the time they are most useful.

The examples in the *Manual* introduce students to a wide variety of their calculator's features. With the aid of calculator or computer technology, many problems encountered in precalculus have multiple methods of solution. We have not tried to cover all possible ways that the calculator can be used to solve particular problems. For example, although the poly command on the TI-85/86 can be used to determine the roots of a polynomial, the guide shows a graphical approach. Students should not feel limited by our suggestions, but should be encouraged to explore other possibilities described in their calculator's manual.

Although the calculator reduces the amount of tedious calculations, and, in the case of the TI-89/92Plus/Voyage 200, tedious algebraic manipulations, it does not, however, relieve the students from the responsibility of thinking about mathematics. Several of the examples in the *Manual* are designed to encourage responsible and thoughtful use of graphing technology. With this in mind, we've included several examples that illustrate limitations of graphing technology.

We would like to acknowledge Hortencia Garcia and Ying Huang of Eastern Connecticut State University for their careful review of the *Graphing Calculator Manual*.

Graphing Calculator Guidebook
for the TI-83Plus
To Accompany
Precalculus: Concepts in Context, 2e

This guide provides background on the TI-83Plus graphing calculator that will be useful for *Precalculus: Concepts in Context, 2e*. It consists of a basic tutorial followed by additional instructions relevant to each chapter of your text. Consult your calculator manual for additional calculator features.

BASIC TUTORIAL

0. *The Keyboard*

Most of the keys on the TI-83Plus access more than one function.
- To access a function written in yellow lettering, press the $\boxed{\text{2nd}}$ key.
   Example: $\boxed{\text{2nd}}$ $\boxed{\text{QUIT}}$ (same key as $\boxed{\text{MODE}}$).
- To access a function written in green, press the $\boxed{\text{ALPHA}}$ key.
   Example: $\boxed{\text{ALPHA}}$ $\boxed{\text{L1}}$ (same key as $\boxed{1}$).

You should familiarize yourself with the locations of the keys that control the cursor. These arrow keys are located on the right-hand side of the calculator (below the $\boxed{\text{TRACE}}$ and $\boxed{\text{GRAPH}}$ keys) and point out in four directions as shown in Figure 1.

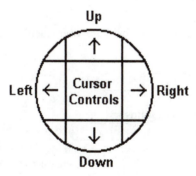

Figure 1. The Arrow Keys

Finally, take a moment to look at the last column of keys on the right side of the calculator. Here you will find the operation keys for addition, subtraction, multiplication, division, and powers. You will also find the $\boxed{\text{CLEAR}}$ and $\boxed{\text{ENTER}}$ keys in this column.

1

TI-83Plus Guide

1. *On, Off, and Contrast*

Turn the calculator on by pressing ⌈ ON ⌉.

You may need to adjust the contrast. Press the ⌈ 2nd ⌉ key followed by holding down the up arrow key to darken or the down arrow key to lighten.

To turn your calculator off, press ⌈ 2nd ⌉ ⌈ OFF ⌉. If you forget, the calculator will automatically turn off after a period of non-use.

2. *Basic Calculations*

When you turn your calculator on, you see the home screen. Your calculations will be displayed on the home screen. If, at any time, you want to clear the home screen, press ⌈ CLEAR ⌉. (If pressing ⌈ CLEAR ⌉ does not work, press ⌈ CLEAR ⌉ a second time or press ⌈ 2nd ⌉ ⌈ QUIT ⌉ followed by ⌈ CLEAR ⌉.) You do not need to clear the home screen after each computation. If not cleared, the home screen will keep a history of several lines of your work.

Work through the following examples to learn about basic computations and techniques on the TI-83Plus.

Example: Compute 3×4.
> After pressing ⌈ 3 ⌉ ⌈ × ⌉ ⌈ 4 ⌉, press ⌈ ENTER ⌉. Note that the original problem, written as $3 * 4$, remains on the left side of the screen and the answer appears on the right.

Example: Compute $3 + 2 \times 6$ and then $(3 + 2) \times 6$.
> After you complete the first calculation, you'll use your calculator's editing features to save keystrokes when entering the second calculation.

> • First, press ⌈ 3 ⌉ ⌈ + ⌉ ⌈ 2 ⌉ ⌈ × ⌉ ⌈ 6 ⌉ ⌈ ENTER ⌉. Do not press ⌈ CLEAR ⌉.

> • Complete the second calculation as follows:
>> Press ⌈ 2nd ⌉ ⌈ ENTRY ⌉ (same key as ⌈ ENTER ⌉) to rewrite the original calculation.
>> Press the left arrow key to move the cursor (a blinking rectangle) over the 3.
>> Press ⌈ 2nd ⌉ ⌈ INS ⌉ (same key as ⌈ DEL ⌉). The cursor turns into a blinking underline. Then press ⌈ (⌉ to insert the parenthesis to the left of 3.
>> Press the right arrow key to move the cursor over the *. Press ⌈ 2nd ⌉ ⌈ INS ⌉ ⌈) ⌉.
>> Press ⌈ ENTER ⌉. You should get 30.

Example: Compute 8^3.
> Press ⌈ 8 ⌉ followed by ⌈ ^ ⌉ ⌈ 3 ⌉ ⌈ ENTER ⌉.

Example: Compute $\sqrt{16}$.

Press 2nd [$\sqrt{}$] (same key as x^2) followed by 1 6 ENTER . Notice that your calculator completed the calculation even though you failed to close the parentheses by pressing) .

Warning! *Make sure you close all parentheses. There should be the same number of right parentheses as left parentheses. Failure to match the parentheses can lead to errors in some calculations.*

Example: Compute $\sqrt{-16}$.

Warning! *The TI-83Plus has two minus keys,* — *and* (-) *, to differentiate between the operation of subtraction (such as $3 - 2 = 1$) and the opposite of the positive number 16, namely -16.*

Press 2nd [$\sqrt{}$] followed by (-) (the key to the left of ENTER) 1 6) ENTER .

OOPS! If your calculator is in Real mode, then you will not be allowed to take the square root of a negative number. Your screen should match the error message in Figure 2. Press 1 *to exit the error message.*

Figure 2. Error Message    Figure 3. Mode Screen

Press MODE . Your mode screen should be similar to the one in Figure 3. Press the down arrow to move the blinking cursor over REAL. Press the right arrow to move the cursor over $a + bi$. Then press ENTER to highlight the new setting. Press 2nd QUIT to return to the home screen.

Now try the example again: press ENTER . This time you should get the imaginary number $4 \cdot i$.

Return your mode setting to REAL. (We will be dealing only with real numbers in this course.)

Example: Find a decimal approximation of $\sqrt{5}$.

Press 2nd $\sqrt{}$ 5) ENTER . In this case, your calculator provides a decimal approximation of $\sqrt{5}$, which is around 2.236. This is not an exact answer (even though the answer on your screen may be precise to 9 decimal places!).

Example: Compute $\sqrt[5]{32}$.

Press 5 for the fifth root and then MATH 5 for $\sqrt[x]{}$. Next, press 3 2 ENTER . Did you get 2?

3. Correcting an Error and Editing

If you make an error in calculation, you may be able to correct the error without reentering the problem. You can also use the editing feature to modify previous calculations. Check out the following two examples.

Example: Make a deliberate error by entering 3 + + 2 and then correct it.
Press ⎡3⎤ ⎡+⎤ ⎡+⎤ ⎡2⎤ ⎡ENTER⎤. To correct the error, press ⎡2⎤ for Goto. The cursor will direct you to the error. Erase one of the plus signs by pressing ⎡DEL⎤, and then press ⎡ENTER⎤. The correct answer to 3 + 2 will appear.

Example: Enter the problem (3 + 2 × 6)/5. Then change it to (3 + 2 × 6)/7.
- Enter the first problem: press ⎡(⎤ ⎡3⎤ ⎡+⎤ ⎡2⎤ ⎡×⎤ ⎡6⎤ ⎡)⎤ ⎡÷⎤ ⎡5⎤ ⎡ENTER⎤.

- Press ⎡2nd⎤ ⎡ENTRY⎤ (same key as ⎡ENTER⎤). Your screen should be similar to Figure 4.

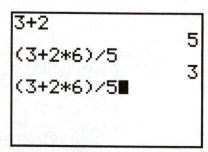

Figure 4. Editing a Calculation

- Press the left arrow key to move the cursor over 5, and then press ⎡7⎤ ⎡ENTER⎤. (Because the cursor was a solid rectangle, you overwrote the 5 when you pressed ⎡7⎤.)

- Your calculator reported the answer as a decimal. If you want the equivalent fraction, press ⎡MATH⎤ ⎡1⎤ for Frac and then ⎡ENTER⎤. You should get $\frac{15}{7}$.

4. Resetting the Memory

After experimenting with settings, you may want to return your calculator's settings to the factory settings. Here's how:
- Press ⎡2nd⎤ ⎡MEM⎤ (same key as ⎡+⎤) ⎡7⎤ for Reset. You have two choices:
  1: All RAM (erases all data and programs)
  2: Defaults (returns all settings to the factory defaults).

- Press ⎡2⎤ ⎡2⎤. The message "Defaults set" should appear on your screen. If you can't read this message, darken the contrast. (Instructions for adjusting the contrast appear in Topic 1, page 2.) Then press ⎡CLEAR⎤ to clear the screen.

4

5. Graphing

Example: Graph $y = x$, $y = x^2$, and $y = x^3$ in the standard viewing window.

You'll tackle this problem in two steps. In step 1, you'll enter the function. In step 2, you'll graph it and then look at the window settings.

Step 1: Enter the three functions as follows.
- Press $\boxed{\text{Y=}}$.

- If you have functions stored in your calculator, use the up or down arrow keys to position the cursor on the same line as the function. Then press $\boxed{\text{CLEAR}}$ to erase the previously stored function.

- Enter the first function to the right of Y1 by pressing $\boxed{\text{X,T,}\theta\text{,}n}$ followed by $\boxed{\text{ENTER}}$.

- Enter the second and third functions to the right of Y2 and Y3, respectively. (Remember to press $\boxed{\wedge}$ to obtain the powers.)

Step 2: Set the standard viewing window and graph. Then check the window settings.
- Press $\boxed{\text{ZOOM}}$ to display the Zoom menu. Then press $\boxed{6}$ to select ZStandard. The graphs of all three functions should appear on your screen.

- To check the window settings, press $\boxed{\text{WINDOW}}$. Your screen should match Figure 5.

```
WINDOW
 Xmin=-10
 Xmax=10
 Xscl=1
 Ymin=-10
 Ymax=10
 Yscl=1
 Xres=1
```

Figure 5. Window Settings

- To exit this menu (or any menu) and return to the home screen, press $\boxed{\text{2nd}}$ $\boxed{\text{QUIT}}$ (same key as $\boxed{\text{MODE}}$).

Do <u>not</u> erase the functions you stored as Y1, Y2 and Y3 until you have completed this tutorial.

Example: Graph $y = x$, $y = x^2$, and $y = x^3$, the functions from the previous example, using the window settings shown in Figure 6. (See the next page.)

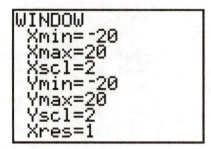
Figure 6. New Window Settings

- Press ⌐WINDOW⌐ to access the Window editor. Change the first setting to − 20 (*remember to press* ⌐(-)⌐ *when you enter* − 20) and press ⌐ENTER⌐. Change the remaining settings to match those in Figure 6.

- Press ⌐GRAPH⌐ to view the graphs in the new window.

Example: Zoom in twice on the graph from the previous example to get a magnified view around a point of intersection. Then use Trace to estimate the coordinates of this point.
- Press ⌐ZOOM⌐ and then ⌐2⌐ for Zoom In. Notice that the center of this first zoom (marked by a blinking pixel) will be (0,0). Press ⌐ENTER⌐ to zoom in on the origin. Your graph should be similar to the one in Figure 7.

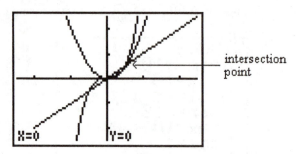
Figure 7. Intersection Point

- Notice the point of intersection to the right of (0,0). (See Figure 7.) Next, you'll zoom in using a center close to this point of intersection. Press the right and up arrow keys to move the cursor (a plus sign) close to this point of intersection. Then press ⌐ENTER⌐. The intersection point should be clearly visible near the center of the screen.

- Press ⌐TRACE⌐. Trace along the first curve by pressing the left and right arrow keys. You should see the cursor (a square with a blinking ×) move along the curve. Press the up or down arrow keys to jump from one curve to another.

- Move the cursor directly on top of the intersection point that lies to the right of the origin. Read the approximate values of the x- and y-coordinates from the bottom of the screen. (The coordinates should be close to $(1,1)$.)

Example: Take a closer look at the graph of $y = x$ without the graphs of the other functions. Then view the graph in a window that uses square scaling.

You'll tackle this problem in three steps. In Step 1, you'll turn off the functions stored in Y2 and Y3. In Step 2, you'll set up the window and view the graph. In Step 3, you'll clear the stored functions.

Step 1: Remove the graphs of $y = x^2$ and $y = x^3$ from the viewing screen without erasing these functions.
- Press $\boxed{\text{Y=}}$ to display the stored functions.

- To unselect Y2, use the up or down arrow key to highlight the equals sign in Y2's formula. Press $\boxed{\text{ENTER}}$ to remove the highlighting over the equals sign.

- Next, move the cursor over the equals sign in Y3's formula. Press $\boxed{\text{ENTER}}$ to remove the highlighting over the equals sign.

Step 2: View the graph of $y = x$ in the standard viewing window and then switch to square scaling.
- Press $\boxed{\text{ZOOM}}$ $\boxed{6}$ to view the graph of $y = x$ in the standard viewing window.

The line should make a 45° angle with the x-axis, but the apparent angle on this screen is less than 45°. (Note that the tick marks on the y-axis are much closer together than on the x-axis.)

- Press $\boxed{\text{ZOOM}}$ $\boxed{5}$ for ZSquare.

Now you should see a true 45° angle. (Note that the tick marks on the x- and y-axes are equally spaced.)

Note: *You can turn the functions for Y2 and Y3 back on by repeating Step 1. This time the process will restore the highlighting over the equals signs in the formulas for Y2 and Y3.*

That's it! You have completed the tutorial. Now practice and experiment on your own with the calculator until you feel comfortable with these basic operations. The remainder of this guide will introduce new techniques and occasionally review techniques as they are needed, chapter by chapter, for your work in *Precalculus: Concepts in Context, 2e.*

7

Chapter-by Chapter Guide

CHAPTER 1

Returning to and Clearing the Home Screen

If you want to exit any screen, press $\boxed{\text{2nd}}$ $\boxed{\text{QUIT}}$ (same key as $\boxed{\text{MODE}}$). You will then return to the home screen. To clear the home screen press $\boxed{\text{CLEAR}}$ once or twice.

Plotting Points

You can use your calculator to plot the Fahrenheit-Celsius data from Lab 1A. Then graph your guess for the formula that relates degrees Fahrenheit to degrees Celsius. This will allow you to check how closely the function specified by your formula follows the pattern of the data.

Example: Plot the data in Table 1 and then overlay the graph of $y = 18x + 85$.
　　　You'll work on this problem in four steps. In Step 1, you'll enter the data; in Step 2, you'll plot these data; in Step 3, you'll add a line to your plot; and in Step 4, you'll turn off the data plot.

Sample Data	
x	y
-2	40
-1	60
1	100
3	140

Table 1. Sample Data

Step 1: Clear lists L1 and L2 and enter the data from Table 1.
　　　The TI-83Plus has six lists, L1 − L6. You'll store the data from the x-column in list L1 and the y-column in list L2. If you have previously deleted these lists, you will need to set them up again. Any previously stored data in these lists will need to be cleared before entering the new data.

- Here's one way to clear lists L1 and L2:
　　　Press $\boxed{\text{STAT}}$ $\boxed{4}$ $\boxed{\text{2nd}}$ $\boxed{\text{L1}}$ (the same key as $\boxed{1}$) $\boxed{,}$ $\boxed{\text{2nd}}$ $\boxed{\text{L2}}$.
　　　Press $\boxed{\text{ENTER}}$. Your calculator will respond with the message "Done."

- Press $\boxed{\text{STAT}}$ $\boxed{1}$ for Edit. Check that you have lists L1 and L2 on your screen. (If necessary, use the left and right arrow keys to scroll from one list to the next.)

　　　If you are missing lists L1 and L2, or they are not in order, press $\boxed{\text{STAT}}$ $\boxed{5}$ for SetUpEditor. Then enter the lists you want to set up, in the order you want them. For example, press $\boxed{\text{2nd}}$ $\boxed{\text{L1}}$, $\boxed{\text{2nd}}$ $\boxed{\text{L2}}$, $\boxed{\text{2nd}}$ $\boxed{\text{L3}}$, $\boxed{\text{2nd}}$ $\boxed{\text{L4}}$, $\boxed{\text{2nd}}$ $\boxed{\text{L5}}$, $\boxed{\text{2nd}}$ $\boxed{\text{L6}}$ $\boxed{\text{ENTER}}$ to set up all six lists. Then press $\boxed{\text{STAT}}$ $\boxed{1}$ to return to the lists.

• Use the arrow keys to highlight the cell directly under L1 as shown in Figure 8.

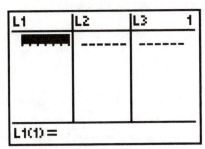

Figure 8. List Screen

• The screen in Figure 8 shows three empty lists, L1 – L3. At the bottom of the screen, L1(1)= indicates that the first entry in L1 is highlighted. Press the right arrow key to highlight the first entry in L2 and observe that the bottom of the screen now displays L2(1)=. Press the left arrow key to return to L1(1).

• Enter the x-data in L1: press $\boxed{\text{(-)}}$ $\boxed{2}$ $\boxed{\text{ENTER}}$ $\boxed{\text{(-)}}$ $\boxed{1}$ $\boxed{\text{ENTER}}$ $\boxed{1}$ $\boxed{\text{ENTER}}$ $\boxed{3}$ $\boxed{\text{ENTER}}$.

• Press the right arrow key to move the cursor to the first cell in list L2. Enter the data from the y-column into L2.

Step 2: Plot the sample data.
• Before plotting the data, clear any previously stored functions. Press $\boxed{\text{Y=}}$, position the cursor on any line with a stored function, and press $\boxed{\text{CLEAR}}$.

• Press $\boxed{\text{2nd}}$ $\boxed{\text{STAT PLOT}}$ (same key as $\boxed{\text{Y=}}$). The menu that appears gives you a choice of three plots. Press $\boxed{1}$ for Plot 1. Your screen should be similar to Figure 9.

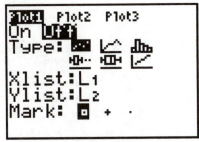

Figure 9. STAT PLOT Menu

• Press $\boxed{\text{ENTER}}$ to highlight On.

9

- Press the down arrow key to move to Type. If not already highlighted, highlight the first choice (the scatterplot) and press ENTER .

- Press the down arrow key to move to Xlist. Press 2nd L1 if a different list is entered for Xlist. If necessary, adjust the selection for Ylist to L2.

- Finally, press the down arrow key to move the cursor to the last line of the menu. Then choose one of the three symbols (square, plus or dot) by highlighting the symbol and pressing ENTER .

- Press ZOOM . Use the down arrow key to scroll down until the number corresponding to ZoomStat is highlighted. Then press ENTER to view the plot. (See Figure 10.) When you use ZoomStat, your calculator automatically selects a window that displays all of your data. If the y-axis appears thick, that's because the setting for Yscl is too small; hence, the tick marks on the y-axis are too close together. You can turn the tick marks off by pressing WINDOW and setting Yscl to 0.

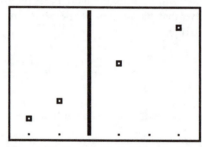

Figure 10. Scatterplot of Sample Data

Step 3: Add the graph of $y = 18x + 85$ to the plot of the sample data.
- Press Y= . Enter the function $y = 18x + 85$ as Y1: press 1 8 X,T,θ,n + 8 5 .

- Press GRAPH .

Step 4: Turn off all Plots as follows.
- Press 2nd STAT PLOT

- Press 4 for PlotsOff and then ENTER . Your calculator will respond "Done."

Warning! *If you don't turn Plot 1 off, your calculator will attempt to plot the data in lists L1 and L2 every time you graph. If you delete or change the data, your calculator will still try to graph Plot 1 the next time you press* GRAPH . *In that case, you will get the error message shown on the next page in Figure 11. So, remember to turn data plots off!*

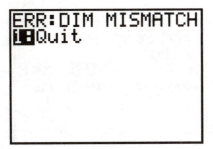

Figure 11. Error Message

Making a Table of Values from a Formula

The TI-83's table feature provides numeric information about a function. In the next example, you'll start with a column of values for the independent variable, x, and use the table feature to calculate the corresponding values for the dependent variable, y.

Example: Use your TI-83Plus to complete Table 2 for the function $y = 2x - 30$.

x	y
2.0	
6.0	
10.0	
14.0	
18.0	
22.0	
26.0	

Table 2. Table of Function Values

Plan of action: First, you'll enter the function into the calculator. Then you'll set up the values for the independent variable, x. Notice that the minimum x-value in the table is 2.0 and that consecutive x-values are separated by increments of 4.0. Using the TABLE command, you'll be able to generate the values for the x- and y-columns.

- Press [Y=]. Clear any previously stored functions and then enter the function $y = 2x - 30$ as Y1.

- Press [2nd] [TblSet] (same key as [WINDOW]) to access the TABLE SETUP screen.

- Change the settings to match those in Figure 12.

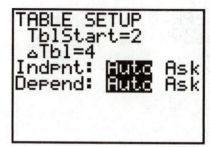

Figure 12. Setting Up a Table

- To view the table, press 2nd TABLE (same key as GRAPH).

  Using the up and down arrows, you can scroll up or down in the table.

  For example, use the down arrow key to find the value of Y1 that corresponds to an x-value of 62. (You should get 94.)

  Now, use the up arrow key to find two x-values in the table between which the dependent variable changes sign. (You should get x-values of 14 and 18.)

- Press 2nd QUIT to exit the table and return to the home screen.

Computing the Value of a Function

Example: Find the value of the function $f(x) = 3x^2 + 6x - 7$ for $x = 5$ and $x = 8$. In other words, find the values of $f(5)$ and $f(8)$.

- Press Y= and erase any previously stored functions. Then enter the formula for $f(x)$ as Y1. Press 2nd QUIT to return to the home screen.

- Calculate Y1(5) as follows:
  Press VARS .
  Use the right arrow key to highlight Y-VARS and press 1 for Function.
  Press 1 to select Y1 and write it on the home screen.
  Now press (5) ENTER . You should get 98.

- Compute Y1(8) as follows. Press 2nd ENTRY (same key as ENTER). Use the left arrow key to move the cursor over 5 and then press 8 ENTER . You should get 233.

Do <u>not</u> erase the formula for $f(x)$ until you have completed the next example.

Example: Find the average rate of change of $f(x) = 3x^2 + 6x - 7$ from $x = 5$ to $x = 7$.

From the previous example, you should already have the formula for $f(x)$ stored as Y1. On the home screen, enter the formula for the average rate of change and press [ENTER]. (See Figure 13.)

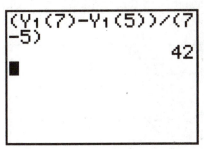

Figure 13. Calculating an Average Rate of Change

Adjusting the Window Settings for Square Scaling

The viewing screen on your calculator is a rectangle. Therefore, if you use the standard window, the tick marks on the y-axis will be closer together than those on the x-axis. For square scaling, you want the distance between 0 and 1 on the x-axis to be the same as the distance between 0 and 1 on the y-axis.

Example: The graph of $f(x) = \sqrt{25 - x^2}$ forms the top half of a circle. Graph $f(x) = \sqrt{25 - x^2}$, first in the standard window, and then switch to square scaling.

You'll do this problem in two steps. In Step 1, you'll enter the function and graph it in the standard viewing window. In Step 2, you'll change to square scaling.

Step 1: Enter $\sqrt{25 - x^2}$ as Y1 and then graph it in the standard viewing window.

- Press [Y=]. Clear any stored functions then enter $\sqrt{25 - x^2}$ as Y1. (Remember there should be both left and right parentheses.)

- Press [ZOOM] [6] for ZStandard.
  Observe the spacing between the tick marks on the x- and y-axes. Notice that the tick marks on the y-axis are closer together than the tick marks on the x-axis. This graph does not appear to be a semicircle.

- Press [WINDOW] to view the window settings of the standard window.

Step 2: Change to square scaling.

- Press [ZOOM] [5] to select ZSquare. When the graph appears on your screen, observe the equal distance between tick marks on the two axes. In this window, the graph is shaped like a semicircle.

- Press [WINDOW] and note the changes in the window settings.

Example: Add the graph of $g(x) = -\sqrt{25 - x^2}$ to the previous graph.

This function is the opposite of the previous function. So, you can define it in terms of Y1. Here's how.

- Press $\boxed{Y=}$. Your screen should match Figure 14. Move the cursor, in this case a solid rectangle, opposite Y2.

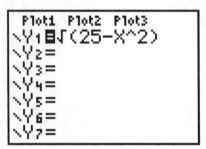

Figure 14. Contents of Y= Editor

- Press $\boxed{(-)}$. Then press $\boxed{VARS}$, use the right arrow to highlight Y-VARS, and press $\boxed{1}$ $\boxed{1}$ to insert Y1 on the screen. Press $\boxed{(}$ $\boxed{X,T,\theta,n}$ $\boxed{)}$.

- To see the graph of both halves of a circle, press $\boxed{GRAPH}$. Your graph should be similar to the one in Figure 15.

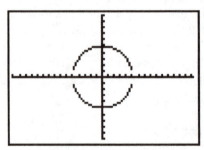

Figure 15. Graph of a Circle

CHAPTER 2

Finding Values of a Recursive Function

Example: Suppose that $P(0) = 2$ and that $P(t + 1) = P(t) + 5$. Use your calculator to find $P(1)$, $P(2)$, $P(3)$, and $P(4)$.

There are several ways to solve this problem. Here's one that uses your calculator's answer feature.

- Enter the value for $P(0)$: press $\boxed{2}$ $\boxed{\text{ENTER}}$.

- Since $P(1) = P(0) + 5$, press $\boxed{+}$ $\boxed{5}$ $\boxed{\text{ENTER}}$ to add 5 to the value of $P(0)$.

- Press $\boxed{\text{ENTER}}$ three more times to find the values of $P(2)$, $P(3)$, and $P(4)$. (Check that your answers match those in Figure 16.)

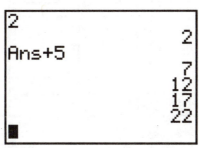

Figure 16. The Answer Feature

Fitting a Line to Data

If, when plotted, your data lie exactly on a line, you can use algebra to determine the equation of the line. However, real data seldom fall precisely on a line. Instead, the plotted data may exhibit a roughly linear pattern. The least squares line (also called the regression line) is a line that statisticians frequently use when describing a linear trend in data.

Example: Use the least squares line to describe the linear pattern in the data displayed in Table 3.

x	y
-3.0	-6.3
-2.0	-2.8
1.2	2.0
2.0	4.1
3.1	5.0
4.2	7.2

Table 3. Data With Linear Trend.

You'll tackle this problem in four steps. In Step 1, you'll enter the data. In Step 2, you'll determine the equation of the least squares line. Then in Step 3, you'll plot the data and graph the line. You'll erase lists X and Y in Step 4.

Step 1: Erase any stored functions and then enter the data.
- Press Y= and erase any stored functions.

- Press STAT   1 to enter the lists.

- Position the cursor so that L1 is highlighted. To form a list named x, press 2nd   INS . Your screen should match the one in Figure 17. Notice that the blinking cursor at the bottom of your screen contains the letter A. That means the ALPHA key has been activated. Press X (same key as STO ▷) and then ENTER .

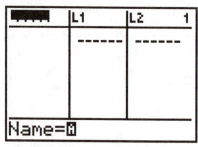

Figure 17. Naming a List

- Press the down arrow key to highlight the first cell in list X and then enter the x-data.

- Use the arrow keys to highlight L1. Press 2nd   INS   Y (same key as 1) to form a new list named Y. Press ENTER .

- Press the down arrow key to highlight the first cell in list Y and enter the y-data.

Step 2: Next, you'll find the equation for the least squares line. Your calculator will determine the values for the slope and intercept of $y = ax + b$.
- Press STAT , use the right arrow key to highlight CALC, and then press 4 (for LinReg(ax+b)). (*If you want to fit a different function, such as an exponential function or a quadratic function, see the note at end of this section.*) Complete the command as follows.
  Press 2nd   List (the same key as STAT), use the down arrow key to highlight X, and press ENTER .
  Press , .
  Press 2nd   List , use the down arrow key to highlight Y, and press ENTER .
  Press ,

16

Press VARS , use the right arrow key to highlight Y-VARS. Press 1 for FUNCTION and then press 1 for Y1. Your command on your screen should match the one in Figure 18. Press ENTER to execute the command.

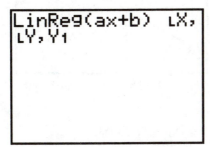

Figure 18. LinReg Command

The output from the LinReg command tells you the values for a and b, the slope and y-intercept, respectively. (If your calculator's diagnostics are turned on, you will also see values for r, Pearson's correlation coefficient, and r^2, the coefficient of determination.)

• Press Y= . Observe that the equation of the least squares line has been stored as Y1.

Step 3: Make a scatterplot of the data and overlay a graph of the least squares line.
• Press 2nd STAT PLOT 1 .

• Press ENTER to turn the plot on. Adjust settings to match those in Figure 19. To enter X and Y, you can press ALPHA X and ALPHA Y , respectively. You can also access X and Y from the stored lists by pressing 2nd LIST .

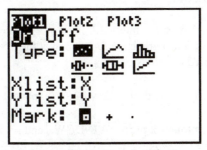

Figure 19. Settings for Plot 1

• Press ZOOM 9 for ZoomStat to see a scatterplot of the data and a graph of the least squares line. Your screen should resemble Figure 20.

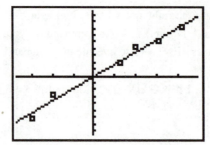

Figure 20. The Least Squares Line

Step 4: Erase lists X and Y.
- Press ⟨2nd⟩ ⟨MEM⟩ ⟨2⟩ for Mem Mgmt/Del.

- Press ⟨4⟩ for List.

- Use the down arrow key to move the triangle so that it points to X. Press ⟨DEL⟩. Then move the triangle so that it points to Y and press ⟨DEL⟩. Lists X and Y have been erased.

Warning! *Remember to turn off this plot. Press* ⟨2nd⟩ ⟨STAT PLOT⟩ ⟨4⟩ *for PlotsOff and then* ⟨ENTER⟩.

Note: *You can adapt the instructions above to fit other functions to data. For example, if you want to fit an exponential function (instead of a linear function), select ExpReg (instead of LinReg) for the Calculation Type. QuadReg will fit a quadratic function, CubicReg will fit a cubic function, SinReg will fit a sinusoidal function, and so forth.*

Graphing an Exponential Function

Example: Graph $p(x) = 200(1.09)^x$ over the interval from $x = 0$ to $x = 10$. Use ZoomFit to choose the window settings for Ymin and Ymax. Then graph the function over the interval from $x = 0$ to $x = 50$
- Press ⟨Y=⟩ and clear any stored functions.

- Enter the formula for $p(x)$ as Y1. (Use the ⟨^⟩ key for the exponent.)

- Change the window settings as follows: press ⟨WINDOW⟩ and set Xmin = 0, Xmax = 10, and Yscl = 0 (to turn off the tick marks on the y-axis).

- Next, let your calculator set the values for Ymin and Ymax. Here's how.
  Press ⟨ZOOM⟩.
  Press the down arrow key until ZoomFit is highlighted, then press ⟨ENTER⟩. Wait a few seconds to view a graph similar to the one in Figure 21.

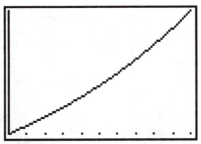

Figure 21. A ZoomFit Window

- Check the window settings: press WINDOW .

- Change the window setting for Xmax to 50. Then press ZOOM , highlight ZoomFit and press ENTER . In this window, the graph should appear to curve upward more steeply than the previous graph.

Finding the Coordinates of Points of Intersection

Example: Find the points where the graphs of $f(x) = -4x + 15$ and $g(x) = -2x^2 + 2x + 12$ intersect. You'll tackle this problem in three steps. In Step 1, you'll graph the functions. You'll find the coordinates of one of the points of intersection in Step 2, and the other in Step 3.

Step 1: Graph the functions in a viewing window that gives a clear view of the points of intersection.
- Enter the two functions into your calculator.

- Adjust the window settings so that you can see both points of intersection. (*Hint:* you might start with the standard viewing window, ZOOM 6 , and then adjust the window settings after looking at the graph.)

Step 2: Approximate the coordinates of one of the points of intersection.
- Press 2nd CALC (same key as TRACE) and then press 5 for intersect. The image on your screen should be similar to Figure 22.

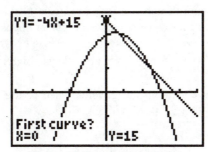

Figure 22. Intersecting Curves

- Your calculator wants to know which curve you want to call the first curve. Try pressing the up and down arrow keys. The cursor will jump back and forth from the line to the parabola. Position the cursor on the line. (We'll designate the line as the first curve since its equation is entered as Y1.) Then press $\boxed{\text{ENTER}}$. The cursor should then jump to the parabola. Press $\boxed{\text{ENTER}}$ to designate the parabola as the second curve.

- Your calculator now asks for a guess. Use the right or left arrow keys to position the cursor (a box with a blinking $\times$) on top of the point of intersection with the smaller x-coordinate. Then press $\boxed{\text{ENTER}}$. Read the coordinates of this intersection point from the bottom of your screen.

Step 3: Find the coordinates of the other intersection point.
Repeat the process outlined in Step 2 to find the coordinates of the other intersection point.

If you have done everything correctly, you will find that the two graphs intersect at approximately (0.634, 12.464) and (2.366, 5.536).

Graphing Piecewise-Defined Functions

Example: Graph the piecewise-defined function $f(x) = \begin{cases} x - 4, & \text{if } x > 4 \\ -x + 4, & \text{if } x \leq 4 \end{cases}$.

The graph of $f(x)$ consists of two half-lines pieced together. You'll want the graph of $y = x - 4$ when x-values are greater than 4 and $y = -x + 4$ when x-values are less than or equal to 4.

You'll complete this example in three steps. In Step 1, you'll enter the functions and in Step 2, you'll piece them together. In Step 3, you'll graph the function.

Step 1: Enter the functions that you want to piece together.
- Press $\boxed{\text{Y=}}$ and clear any previously stored functions. Then enter $x - 4$ as Y1 and $-x + 4$ as Y2.

- Press $\boxed{\text{ZOOM}}$ $\boxed{6}$. Your graph should look like a cross: $\times$.

Step 2: Piece together the functions entered in Step 1 to create $f(x)$.
- Press $\boxed{\text{Y=}}$. Use the down arrow key to move the cursor opposite Y3.

- To insert Y1: press $\boxed{\text{VARS}}$, use the right arrow key to highlight Y-VARS, then press $\boxed{1}$ $\boxed{1}$.

- Press $\boxed{\times}$. Enter the condition that governs when to use Y1: press $\boxed{(}$ $\boxed{\text{X,T,}\theta\text{,}n}$ $\boxed{\text{2nd}}$ $\boxed{\text{TEST}}$ (same key as $\boxed{\text{MATH}}$) $\boxed{3}$ $\boxed{4}$ $\boxed{)}$.

- Press $\boxed{+}$.

- To insert Y2: press $\boxed{\text{VARS}}$, use the right arrow key to highlight Y-VARS, then press $\boxed{1}$ $\boxed{2}$.

- Press $\boxed{\times}$. Enter the condition that governs when to use Y2: press $\boxed{(}$ $\boxed{\text{X,T,}\theta,n}$ $\boxed{\text{2nd}}$ $\boxed{\text{TEST}}$ $\boxed{6}$ $\boxed{4}$ $\boxed{)}$ $\boxed{\text{ENTER}}$. When you have completed this step, your screen should be similar to Figure 23.

```
Plot1  Plot2  Plot3
\Y1 ◻X-4
\Y2 ◻ -X+4
\Y3 ◻Y1*(X>4)+Y2*
(X≤4)
\Y4=
\Y5=
\Y6=
```

Figure 23. Entering the Formula For $f(x)$

Here's how your calculator interprets the information you've just entered as a piecewise-defined function. The calculator assigns the expression $(x > 4)$ the value 1 when the inequality is true (in other words, when the input variable, x, is greater than 4). In this case, the inequality $x \leq 4$ is false, so the calculator sets the expression $x \leq 4$ equal to 0. Thus, for $x > 4$, the function Y3 is equivalent to:

$$Y3 = (x - 4)(1) + (-x + 4)(0) = x - 4.$$

And when $x \leq 4$, the function Y3 is equivalent to:

$$Y3 = (x - 4)(0) + (-x + 4)(1) = -x + 4.$$

Step 3: Graph $f(x)$.
- Unselect Y1 and Y2:
    Position the cursor over Y1's equals sign and press $\boxed{\text{ENTER}}$.
    Then position the cursor over Y2's equals sign and press $\boxed{\text{ENTER}}$.

- Now, press $\boxed{\text{GRAPH}}$. The graph of $f(x)$ should look V-shaped.

Example: Graph $g(x) = \begin{cases} -2x, & \text{if } x < 1 \\ -2 + 3(x - 1), & \text{if } 1 \leq x < 3 \\ 4, & \text{if } x \geq 3 \end{cases}$

In the previous example, you entered separate functions and then pieced them together to make a new single function. This time, you'll enter the three-piece function as Y1. There is one problem: your calculator is not sophisticated enough to understand the condition $1 \leq x < 3$. Instead you will have to enter the equivalent condition $1 \leq x$ and $x < 3$.

- Press $\boxed{\text{Y=}}$ and erase any stored functions.

21

- Enter $g(x)$ as Y1:

  Press [(-)] [2] [X,T,θ,n].

  Press [×] [(] [X,T,θ,n] [2nd] [TEST] [5] [1] [)].

  Press [+].

  Press [(] [(-)] [2] [+] [3] [(] [X,T,θ,n] [−] [1] [)] [)].

  Press [×] [(] [1] [2nd] [TEST] [6] [X,T,θ,n] [2nd] [TEST], use the right arrow key to highlight LOGIC, press [1] for *and*, then press [X,T,θ,n] [2nd] [TEST] [5] [3] [)].

  Press [+].

  Press [4] [×] [(] [X,T,θ,n] [2nd] [TEST] [4] [3] [)].

  Press [ENTER]. (See Figure 24.)

- Press [ZOOM] [6] to graph $g(x)$. If you entered the function correctly, your graph should match the one shown in Figure 25.

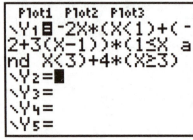

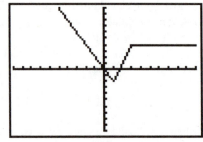

Figure 24. The Formula for $g(x)$ Figure 25. The Graph of $g(x)$

Example: Graph the greatest integer function, $[[x]]$ in the standard viewing window. (See Chapter 2, Exercise 46 in *Precalculus: Concepts in Context, 2e* for a definition of this function.)

The greatest integer function is one of your calculator's built-in functions. Its calculator name is int(x). You can access it from the catalog.

- Press [Y=] and clear any stored functions. Position the cursor opposite Y1.

- To enter the function, press [2nd] [CATALOG] (same key as [0]) and then press [I] (same key as [x^2]). Press the down arrow key until the triangle points to **int(**. Then press [ENTER] [X,T,θ,n] [)] [ENTER].

- Press [ZOOM] [6] to view the graph. Your graph should resemble Figure 26.

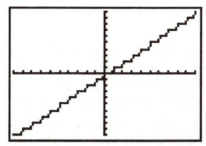

Figure 26. Graph of the Greatest Integer Function

The vertical line segments that appear on your screen cannot be part of the graph of a function. (Why?) For example, the nearly vertical line segment at $x = 3$ is due to the fact that your graphing calculator plotted a point on the left side of $x = 3$, and a point on the right of $x = 3$, and then connected the points with a straight line segment. One solution to this problem is to change the mode from Connected to Dot. That's what you'll do next.

- Press $\boxed{\text{MODE}}$. Move the cursor over Dot (the second setting on the fifth line) and press $\boxed{\text{ENTER}}$. Now press $\boxed{\text{Graph.}}$. The nearly vertical line segments should now be gone.

- Press $\boxed{\text{MODE}}$ and return your mode settings to Connected. Press $\boxed{\text{2nd}}$ $\boxed{\text{QUIT}}$ to return to the home screen.

CHAPTER 3

In Chapter 3 you will be investigating the effect that certain algebraic modifications, such as adding a constant to the input variable, have on the graph of a function. You'll want to experiment using several different functions. We've provided some functions and algebraic modifications that you might want to consider particularly while working on Lab 3, Graph Trek.

Using Parentheses

Warning! *When you want to apply a function to an expression, you must enclose the entire expression in parentheses. For built-in functions (such as the square root or the sine function or the greatest integer function), the TI-83Plus will automatically insert the left parenthesis and you will need to add the right parenthesis.*

Example: Graph $y = \sqrt{x + 2}$ in the window $[-5, 5] \times [-5, 5]$.
- Press $\boxed{\text{Y=}}$ and erase any previously stored functions. Then enter $y = \sqrt{x + 2}$ by pressing $\boxed{\text{2nd}}$ $\boxed{\sqrt{}}$ $\boxed{\text{X,T,}\theta,n}$ $\boxed{+}$ $\boxed{2}$ $\boxed{)}$. (Note: The TI-83Plus automatically inserts the left parenthesis when you press the square root key.)

- Now, press $\boxed{\text{WINDOW}}$ and set Xmin = -5, Xmax = 5, Ymin = -5 and Ymax = 5. Press $\boxed{\text{GRAPH}}$ to graph this function in the $[-5, 5] \times [-5, 5]$ window. Your graph should resemble the one in Figure 27.

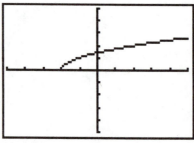

Figure 27. A Member of the Square Root Family

Graphing Functions Involving Absolute Value

Example: Graph $f(x) = |x|$. Then add the graph of $g(x) = \frac{|x|}{x}$. Use a $[-5, 5] \times [-5, 5]$ window. You can access the absolute value function from the catalog.

- Press $\boxed{\text{Y=}}$ and clear any stored functions.

- Enter $y = |x|$ as Y1: press $\boxed{\text{2nd}}$ $\boxed{\text{CATALOG}}$. If necessary, use the arrow keys to move the triangle so that it points to **abs(** and press $\boxed{\text{ENTER}}$. Then press $\boxed{\text{X,T,}\theta,n}$ $\boxed{)}$ $\boxed{\text{ENTER}}$.

24

- If you worked through the previous example, you already have the correct window settings. (Otherwise, press $\boxed{\text{WINDOW}}$ and adjust the window settings.) Press $\boxed{\text{GRAPH}}$ to view the V-shaped graph of the absolute value function.

- Next, enter the formula for $g(x)$ as Y2:
  Press $\boxed{\text{Y=}}$ and position the cursor opposite Y2.
  Press $\boxed{\text{2nd}}$ $\boxed{\text{CATALOG}}$ $\boxed{\text{ENTER}}$ for **abs(**.
  Then press $\boxed{\text{X,T,}\theta\text{,}n}$ $\boxed{\text{)}}$ $\boxed{\div}$ $\boxed{\text{X,T,}\theta\text{,}n}$.

- Press $\boxed{\text{GRAPH}}$ to view the two graphs.

Note: *The function $g(x) = \frac{|x|}{x}$ tells you the sign of the input. The value of $g(x)$ is -1 for negative inputs and $+1$ for positive inputs. It is undefined at zero.*

Do <u>not</u> clear Y1 and Y2 from your calculator until you have completed the next example.

Example: Use your calculator to evaluate $f(x) = |x|$ and $g(x) = \frac{|x|}{x}$ at $x = 0$.

From the previous example, you should already have $|x|$ and $\frac{|x|}{x}$ stored as Y1 and Y2, respectively. Begin these calculations from the home screen.

- First, evaluate $f(0)$:
  Press $\boxed{\text{VARS}}$, highlight Y-VARS, and then press $\boxed{1}$ $\boxed{1}$ for Y1.
  Press $\boxed{\text{(}}$ $\boxed{0}$ $\boxed{\text{)}}$ $\boxed{\text{ENTER}}$. You should get 0.

- Next, evaluate $g(0)$:
  Press $\boxed{\text{2nd}}$ $\boxed{\text{ENTRY}}$ to bring back the previous calculation for editing.
  Use the left arrow key to move the cursor over Y.
  Press $\boxed{\text{VARS}}$, highlight Y-VARS, then press $\boxed{1}$ $\boxed{2}$ for Y2.
  Press $\boxed{\text{ENTER}}$. Do you see why $g(x)$ is not defined at $x = 0$?

Note: *If you modify g(x) by setting g(0) = 0, then you get the sign function. The function sign(x) is defined by the following piecewise formula:*

$$\text{sign}(x) = \begin{cases} \frac{|x|}{x}, & \text{if } x \neq 0 \\ 0, & \text{if } x = 0 \end{cases}.$$

The sign function is not to be confused with the sine function, the next topic.

Graphing the Sine Function

Example: Graph $y = \sin(x)$ in the trigonometric viewing window. You will learn more about this function in Chapter 6.

- First check that your calculator is in Radian Mode. Press MODE. If Radian is not highlighted, move the cursor over Radian and press ENTER.

- Press Y= and erase any stored functions. Then enter $y = \sin(x)$ for Y1 as follows: press SIN X,T,θ,n) ENTER.

- Press ZOOM 7 to view the graph of $y = \sin(x)$ in the trig viewing window.

Graphing a Family of Functions

Using your calculator's list capabilities, you can substitute each value in a given list for a constant in an algebraic formula. This feature allows you to graph an entire family of functions quickly. On the TI-83Plus you specify a list by enclosing the members of the list in brackets: { }.

Example: Graph the family of quadratic functions $y = (x+1)^2$, $y = (x+2)^2$, and $y = (x+3)^2$ in the window $[-5, 5] \times [-1, 10]$.
- Press WINDOW and adjust the settings for a $[-5, 5] \times [-1, 10]$ window.

- Enter the three functions by specifying the constants 1, 2, and 3 in a list as follows: Press Y= and clear any stored functions. Position the cursor opposite Y1 and press (X,T,θ,n + 2nd { 1 , 2 , 3 2nd }) ^ 2.

- Press GRAPH and watch as the three functions are graphed one after the other.

Composing Functions

Example: Suppose $f(x) = 2x - 7$ and $g(x) = 5x^2$. Find the value of $f(g(5))$.
- Press Y= and erase any previously stored functions.

- Enter the formula for $f(x)$ as Y1 and the formula for $g(x)$ as Y2. Then press 2nd QUIT to return to the home screen.

- Enter the composition as follows:
  Press VARS.
  Use the right arrow key to highlight Y-VARS, press 1 1 for Y1.
  Press (VARS, use the right arrow key to highlight Y-VARS, and press 1 2 for Y2.
  Press (5)) ENTER.

The value of the composition at $x = 5$ should be 243.

Do <u>not</u> erase $f(x)$ and $g(x)$ until after you have completed the next example.

Example: Given $f(x) = 2x - 7$ and $g(x) = 5x^2$. Graph $f(g(x))$ in a window that shows the key features of the graph.

From the previous example you should have $f(x)$ stored as Y1 and $g(x)$ stored as Y2. First, you'll form the composition and store it as Y3. Then you'll graph it.

- Press ⬚ Y= . Move the cursor opposite Y3.

- Enter the composition:
  Press ⬚ VARS ⬚, highlight Y-VARS, press ⬚ 1 ⬚⬚ 1 ⬚ for Y.
  Press ⬚ (⬚.
  Press ⬚ VARS ⬚, highlight Y-VARS, and then press ⬚ 1 ⬚⬚ 2 ⬚ for Y2.
  Press ⬚ (⬚⬚ X,T,θ,n ⬚⬚) ⬚⬚) ⬚.

- Unselect the functions stored as Y1 and Y2:
  Position the cursor over the Y1's equals and press ⬚ ENTER ⬚ to remove the highlighting.
  Repeat the process for Y2.

- Adjust your window settings for a $[-5, 5] \times [-10, 10]$ window. Then press ⬚ GRAPH ⬚. The graph should look like a parabola.

CHAPTER 4

Graphing Exponential Functions Involving e

There are two bases for exponential functions that are so common that they have their own function keys on the TI-83Plus: $\boxed{e^x}$ and $\boxed{10^x}$.

Example: Graph $f(x) = e^x$ in the window $[-3, 3] \times [-1, 12]$.
- Press $\boxed{\text{Y=}}$ and clear any stored functions. Then enter $f(x)$ as Y1: press $\boxed{\text{2nd}}$ $\boxed{e^x}$ (same key as $\boxed{\text{LN}}$) $\boxed{\text{X,T,}\theta\text{,n}}$ $\boxed{)}$.

- Press $\boxed{\text{WINDOW}}$ and adjust the settings for a $[-3, 3] \times [-1, 12]$ window. Then press $\boxed{\text{GRAPH}}$ to view the graph.

Example: Graph $g(x) = e^{-\frac{x}{2}}$ by editing the function from the previous example.
- Press $\boxed{\text{Y=}}$ to open the Y= editor.

- Use the right arrow key to move the cursor over the X. Press $\boxed{\text{2nd}}$ $\boxed{\text{INS}}$ $\boxed{\text{(-)}}$. Then use the right arrow key to move the cursor over the right parenthesis and press $\boxed{\text{2nd}}$ $\boxed{\text{INS}}$ $\boxed{\div}$ $\boxed{2}$.

- Press $\boxed{\text{GRAPH}}$ to view the function in the same window as the previous example.

Do <u>not</u> erase $g(x)$ until you have worked through the next topic, *Turning Off the Axes*.

Turning Off the Axes

Sometimes it is helpful to view a graph without the presence of the x and y axes. In the last example, you may have noticed that the graph of $g(x)$ disappeared into the x-axis. Did the graph, in reality, disappear? In the next example, you'll turn off the axes and find out.

Example: View the graph of $y = e^{-\frac{x}{2}}$ in the window $[-3, 3] \times [-1, 12]$ with the axes turned off.
- To turn off the axes, press $\boxed{\text{2nd}}$ $\boxed{\text{FORMAT}}$ (same key as $\boxed{\text{ZOOM}}$), use the arrow keys to highlight AxesOff, and press $\boxed{\text{ENTER}}$.

- Press $\boxed{\text{GRAPH}}$.

- Press $\boxed{\text{2nd}}$ $\boxed{\text{FORMAT}}$ and turn the axes back on.

Graphing Logarithmic Functions

The natural logarithmic function (base e), $\ln(x)$, and the common logarithmic function (base 10), $\log(x)$, have their own function keys, $\boxed{\text{LN}}$ and $\boxed{\text{LOG}}$, respectively. Logarithmic functions of other bases can be graphed by dividing $\ln(x)$ or $\log(x)$ by the appropriate scaling factor.

Example: Graph $h(x) = \ln(x)$ and $g(x) = \log(x)$ using settings for a $[-1, 12] \times [-2, 3]$ window.
- Press $\boxed{Y=}$ and clear any previously stored functions.

- Enter $h(x)$ as Y1: press $\boxed{\text{LN}}$ $\boxed{\text{X,T,}\theta\text{,n}}$ $\boxed{)}$ $\boxed{\text{ENTER}}$.

- Enter $g(x)$ as Y2 : press $\boxed{\text{LOG}}$ $\boxed{\text{X,T,}\theta\text{,n}}$ $\boxed{)}$.

- Press $\boxed{\text{WINDOW}}$ and adjust the settings for a $[-1, 12] \times [-2, 3]$ window. Then press $\boxed{\text{GRAPH}}$ to view the two graphs on the same screen.

Approximating Instantaneous Rates of Change

You can approximate the instantaneous rate of change of a function $f(x)$ at $x = a$ by evaluating the average rate of change of the function over a small interval containing a. For example, you might choose a symmetric interval about a of the form $(a - h, a + h)$ for some small value of h. Using this interval, your approximation of the instantaneous rate of change would be

$$\frac{f(a + h) - f(a - h)}{(a + h) - (a - h)} = \frac{f(a + h) - f(a - h)}{2h}.$$

This is called the central difference quotient, which can be computed using the command **nDeriv.** The syntax of the command is shown below:

$$\text{nDeriv(function, variable, value, } h)$$

Example: Use the central difference quotient to approximate the instantaneous rate of change of the function $f(x) = e^{2x}$ at $x = 0$, 1, and 2. Use $h = 0.01$.
- Press $\boxed{Y=}$ and clear any previously stored functions.

- Enter e^{2x} for Y1. (Be sure to enclose the exponent in parentheses.) Then press $\boxed{\text{2nd}}$ $\boxed{\text{QUIT}}$ to return to the home screen.

- Press $\boxed{\text{MATH}}$ and then use the down arrow key to scroll down to **nDeriv(.** Press $\boxed{\text{ENTER}}$.

- Complete the command.
  Press $\boxed{\text{VARS}}$, use the right arrow key to highlight Y-VARS, and then press $\boxed{1}$ $\boxed{1}$ for Y1.
  Press $\boxed{(}$ $\boxed{\text{X,T,}\theta\text{,n}}$ $\boxed{)}$ $\boxed{,}$ $\boxed{\text{X,T,}\theta\text{,n}}$ $\boxed{,}$ $\boxed{0}$ $\boxed{,}$ $\boxed{.}$ $\boxed{0}$ $\boxed{1}$ $\boxed{)}$ $\boxed{\text{ENTER}}$.

Your screen should be similar to Figure 28.

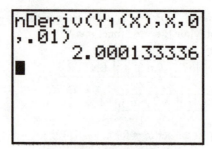

Figure 28. Using nDerive

• Edit the previous command to approximate the rate of change at $x = 1$. Here's how.
Press $\boxed{\text{2nd}}$ $\boxed{\text{ENTRY}}$.
Change the value 0 to 1 and press $\boxed{\text{ENTER}}$.
Repeat the process again, this time changing the value 1 to 2.

You should get approximately 14.8 and 109.2, respectively.

CHAPTER 5

Approximating the Zeros of a Polynomial Function

Example: Find the zeros of $f(x) = 6x^3 - 49x^2 + 50x + 168$. In other words, find the x-values that make the function zero.

To find the zeros of a polynomial function, locate the x-intercepts of its graph.

- Press $\boxed{Y=}$ and clear any previously stored functions. Enter the formula for $f(x)$ as Y1.

- Press $\boxed{ZOOM}$ $\boxed{6}$ to view the graph in the standard viewing window. Notice that this is not a good window for seeing the key features of this function. You'll need to adjust the window settings so that you can clearly see all turning points and x- and y-intercepts: press $\boxed{WINDOW}$ and change the window settings to match those in Figure 29. Then press $\boxed{GRAPH}$. Your graph should look like the one in Figure 29.

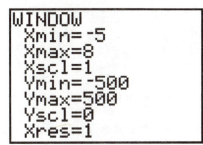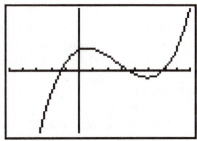

Figure 29. Window Settings and Graph of $f(x)$

Notice that the graph crosses the x-axis in three places. First find the smallest x-intercept (the one that is negative).

- Press $\boxed{2nd}$ $\boxed{CALC}$ (same key as $\boxed{TRACE}$) $\boxed{2}$ (for zero).

- You will need to specify a small interval around the negative x-intercept and then a guess.

  Use the left arrow key to position the cursor (a box with a blinking $\times$) to the left of this x-intercept, and press $\boxed{ENTER}$.

  Use the right arrow key to position the cursor slightly to the right of this x-intercept and press $\boxed{ENTER}$.

  Use the right or left arrow keys to position the cursor close to this x-intercept for your guess. (Your screen should be similar to Figure 30.) Press $\boxed{ENTER}$ and read the x-value from the bottom of your screen.

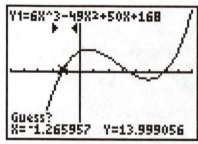

Figure 30. Finding the Zeros of $f(x)$

• Repeat the process to find the values of the two positive zeros.

The zeros of $f(x)$ are $-\frac{4}{3}$ (≈ -1.33), 6, and 3.5.

Finding Local Maxima or Minima

Example: Estimate the local maximum and local minimum of $f(x) = x^3 - 4x^2 + 2x - 4$.
 You'll complete this example in three steps. In Step 1, you'll graph $f(x)$. In Step 2, you'll find the local minimum. In Step 3, you'll find the local maximum.

Step 1: Graph $f(x)$ using a window that gives you a clear view of the two turning points (one peak and one valley) of the graph.
 • Press $\boxed{\text{Y=}}$ and clear any previously stored functions. Enter the formula for $f(x)$ as Y1.

 • Experiment with window settings until you find settings that show both turning points. The graph in Figure 31 shows one example.

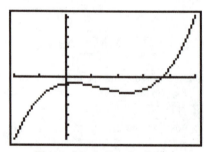

Figure 31. A Graph of $f(x)$

Step 2: Approximate the coordinates of the turning point associated with the local maximum (the y-coordinate of the peak on the graph) as follows:
 • Press $\boxed{\text{2nd}}$ $\boxed{\text{CALC}}$ and then $\boxed{4}$ for maximum.

 • You'll have to specify a narrow interval about the turning point associated with the local maximum (peak). Press the left arrow key to move the cursor slightly to the left of the peak and then press

ENTER . Next, press the right arrow key to position the cursor slightly to the right of the turning point associated with the maximum and press ENTER .

- Next, you'll provide a guess for the turning point: using the right or left arrow keys, move the cursor close to the turning point associated with the maximum. (See Figure 32.) Press ENTER .

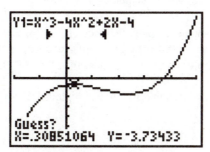

Figure 32. Guess for Local Maximum

- Read off the coordinates of this turning point at the bottom of your screen. The local maximum (y-coordinate of this turning point) should be approximately -3.73.

Step 3: Approximate the coordinates of the turning point associated with the local minimum (the y-coordinate of the valley on the graph).

- Press 2nd CALC 3 for minimum.

- Adapt the instructions for Step 2 to find the coordinates of the turning point associated with the local minimum (the valley). You should get a value for y that is close to -8.42.

Approximating Instantaneous Rates of Change

You can approximate the instantaneous rate of change of a function $f(x)$ at $x = a$ by evaluating the average rate of change of the function over a small interval containing a. For example, you might choose a symmetric interval about a of the form $(a - h, a + h)$ for some small value of h. Using this interval, your approximation of the instantaneous rate of change would be

$$\frac{f(a+h) - f(a-h)}{(a+h) - (a-h)} = \frac{f(a+h) - f(a-h)}{2h}.$$

This is called the central difference quotient.

Example: Use the central difference quotient to approximate the instantaneous rate of change of the function $f(x) = x^3 - 4x^2 + 2x - 4$ at $x = 1$ and $x = 3$. Use $h = .01$.

The formula for the central difference quotient using $a = 1$ and $h = .01$ is $\frac{f(1+.01) - f(1-.01)}{2(.01)}$. You'll do this problem in three steps. In Step 1, you'll enter the function. In Steps 2 and 3, you'll approximate the instantaneous rate of change of $f(x)$ at $x = 1$ and $x = 3$, respectively.

Step 1: Enter $f(x)$ as Y1 and return to the home screen.
- Press $\boxed{\text{Y=}}$. If you completed the previous example, the formula for $f(x)$ may already be stored as Y1. If not, enter the formula for $f(x)$ as Y1.

- Press $\boxed{\text{2nd}}$ $\boxed{\text{QUIT}}$ to return to the home screen.

Step 2: Approximate the instantaneous rate of change of $f(x)$ at $x = 1$.
Enter the formula for the central difference quotient using $x = 1$ and $h = .01$. Make sure you enter the entire formula before pressing enter. Here are the keystrokes.

- Press $\boxed{\text{(}}$ $\boxed{\text{VARS}}$, use the right arrow key to highlight Y-Vars, then press $\boxed{1}$ $\boxed{1}$.

- Press $\boxed{\text{(}}$ $\boxed{1}$ $\boxed{+}$ $\boxed{.}$ $\boxed{0}$ $\boxed{1}$ $\boxed{)}$ $\boxed{-}$.

- Press $\boxed{\text{VARS}}$, use the right arrow key to highlight Y-Vars, then press $\boxed{1}$ $\boxed{1}$.

- Press $\boxed{\text{(}}$ $\boxed{1}$ $\boxed{-}$ $\boxed{.}$ $\boxed{0}$ $\boxed{1}$ $\boxed{)}$ $\boxed{)}$.

- Press $\boxed{\div}$ $\boxed{\text{(}}$ $\boxed{2}$ $\boxed{\times}$ $\boxed{.}$ $\boxed{0}$ $\boxed{1}$ $\boxed{)}$

- Press $\boxed{\text{ENTER}}$. You should get approximately -3.0.

Step 3: Approximate the instantaneous rate of change of $f(x)$ at $x = 3$ by editing work from the previous step.
- Press $\boxed{\text{2nd}}$ $\boxed{\text{ENTRY}}$.

- Change the two x-values of 1 to 3. Then press $\boxed{\text{ENTER}}$. The result is approximately 5.0.

In this example, you approximated instantaneous rates of change directly from a formula. You might prefer using your calculator's **nDeriv** command instead. For more information, check ₹ out *Approximating Instantaneous Rates of Change* in Chapter 4, pages 29 and 30.

Graphing a Rational Function

Warning! *If the numerator or the denominator of a rational function consists of more than one term, you must enclose it in parentheses when you enter it into your calculator.*

Example: Graph $r(x) = \dfrac{x^2 - 1}{x - 3}$ in the standard viewing window. Then adjust window settings to display the key graphical features of this function.
- Press $\boxed{\text{Y=}}$ and erase any previously stored functions.

- Enter $r(x)$ as Y1: press $\boxed{\text{(}}$ $\boxed{\text{X,T,}\theta\text{,}n}$ $\boxed{\wedge}$ $\boxed{2}$ $\boxed{-}$ $\boxed{1}$ $\boxed{)}$ $\boxed{\div}$ $\boxed{\text{(}}$ $\boxed{\text{X,T,}\theta\text{,}n}$ $\boxed{-}$ $\boxed{3}$ $\boxed{)}$.

• Press ZOOM 6 to view the graph in the standard window. Your graph should look similar to the one in Figure 33.

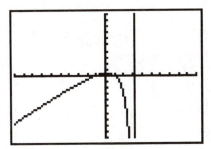

Figure 33. Graph of $r(x)$

The vertical line in the graph shown in Figure 33 indicates that this function has a vertical asymptote at $x = 3$. Remember, this line is not part of the graph of the function. Furthermore, because the domain of this function is all real numbers except $x = 3$, there is a branch of this function's graph that lies to the right of the line $x = 3$. In order to observe this branch, you will have to adjust the window setting for Ymax.

• Press WINDOW . Change the setting for Ymax to 20. Then press GRAPH . You should now see both branches of the graph along with a nearly vertical line that is not part of the graph of this function.

• To remove the vertical line, change from Connected to Dot mode: press MODE , use the arrow keys to highlight Dot, and press ENTER . Now press Graph. . The nearly vertical line should be gone.

• Return the mode to Connected for the next Example.

Do not clear this function from your calculator's memory until after you have completed the next topic, *Zooming Out*.

Zooming Out

Example: Graph the function $r(x) = \dfrac{x^2 - 1}{x - 3}$ in the standard viewing window and then zoom out by a factor of four several times.

For this example, observe what happens to the appearance of the graph of $r(x)$ as we "back away" by increasing the width and height of the viewing window. From the previous example, you should already have the formula for $r(x)$ stored as Y1.

• Press ZOOM 6 to graph the function in the standard viewing window. Then press WINDOW to access the window settings. Set XScl and YScl equal to 0. (This turns off the tick marks that appear on the axes. If you skip this step, the axis will get crowded with tick marks when you zoom out.) Then press GRAPH to return to the graph.

- Check the settings for the zoom factor. Press $\boxed{\text{ZOOM}}$, use the right arrow key to highlight MEMORY and then press $\boxed{4}$ for SetFactors. Your screen should match the one in Figure 34. (If it doesn't, change the settings for XFact and YFact to 4.) Press $\boxed{\text{GRAPH}}$ to return to the graph.

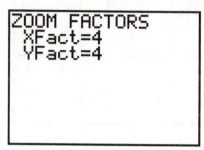

Figure 34. Settings for Zoom Factors

- Press $\boxed{\text{ZOOM}}$ $\boxed{3}$ for Zoom Out. A blinking pixel marks the center of the zoom. (If you want to change the zoom center, use the arrow keys to move the cursor to a new location.) Press $\boxed{\text{ENTER}}$ to view the graph over wider x- and y-intervals.

- Press $\boxed{\text{ENTER}}$ again to zoom out a second time. (Compare your zoomed out graphs with those in Figure 35.)

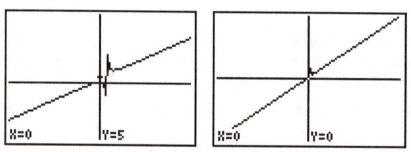

Figure 35. Zooming Out Twice on Figure 33

- Now, press $\boxed{\text{WINDOW}}$ to observe the effect on the window settings of zooming out twice by a factor of 4.

The default setting for Zoom Out widens both the x- and y-intervals by a factor of 4 each time that it is applied. In the previous example, you zoomed out twice. Therefore, the x- and y-intervals are 16 times wider than they were before you zoomed out.

In the next example, you'll change the default zoom settings in order to observe a graph of a function that begins to act like its horizontal asymptote.

Example: Graph $q(x) = \dfrac{5x^2 + 20x - 105}{2x^2 + 2x - 60}$ in the standard viewing window.

• Press ⌈Y=⌉, erase any stored functions and then enter $q(x)$ as Y1. Press ⌈ZOOM⌉ ⌈6⌉. If you have entered the function correctly, your graph should look like the one in Figure 36.

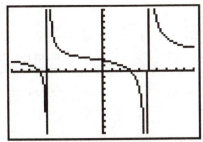

Figure 36. Graph of $q(x)$

Based on this graph, you may suspect that the function $q(x)$ has a horizontal asymptote but this is not at all obvious. Let's observe the function's graph over increasingly wide x-intervals to see if it begins to behave like a horizontal line. Instructions on how to zoom out in the horizontal direction follow.

• Press ⌈WINDOW⌉ and set XScl to 0.

• Press ⌈ZOOM⌉, use the right arrow key to highlight MEMORY, and press ⌈4⌉. Leave the setting for XFact at 4; change the setting for YFact from 4 to 1. Press ⌈GRAPH⌉ to return to the graph.

• Press ⌈ZOOM⌉ ⌈3⌉ ⌈ENTER⌉ to widen the x-interval by a factor of four. Press ⌈ENTER⌉ several more times to continue widening the x-interval. Your graph should begin to resemble its horizontal asymptote $y = 2.5$.

CHAPTER 6

Graphing Trigonometric Functions

Three of the six basic trigonometric functions are built-in functions on the TI-83Plus: sine, $\boxed{\text{SIN}}$, cosine $\boxed{\text{COS}}$, and tangent $\boxed{\text{TAN}}$. Before graphing any of these functions, you should first check that your calculator is set to radian mode: press $\boxed{\text{MODE}}$ and check that Radian is highlighted. (If Degree is highlighted, use the arrow keys to position the cursor over Radian and press $\boxed{\text{ENTER}}$.) Press $\boxed{\text{2nd}}$ $\boxed{\text{QUIT}}$ to return to the home screen.

Example: Graph $y = \sin(x)$ and $y = \csc(x)$ in the trigonometric viewing window.
- Press $\boxed{\text{Y=}}$ and erase any stored functions. Then enter the function $\sin(x)$ as Y1 by pressing $\boxed{\text{SIN}}$ $\boxed{\text{X,T,}\theta\text{,n}}$ $\boxed{)}$ $\boxed{\text{ENTER}}$.

Warning! *Since $\csc(x)$ is defined as $\frac{1}{\sin(x)}$, you will raise sin(x) to the -1 power using the reciprocal key $\boxed{x^{-1}}$. You could also divide 1 by sin(x). However, you <u>cannot</u> use the $\boxed{\text{SIN}^{-1}}$ key which is reserved for the inverse sine function.*

- Enter $\csc(x)$ as Y2: Press $\boxed{(}$ $\boxed{\text{SIN}}$ $\boxed{\text{X,T,}\theta\text{,n}}$ $\boxed{)}$ $\boxed{)}$ $\boxed{x^{-1}}$.

- To graph these functions in the trig viewing window, press $\boxed{\text{ZOOM}}$ $\boxed{7}$ for ZTrig. Your graph should be similar to the one in Figure 37.

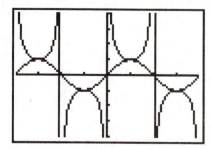

Figure 37. Graphs of $\sin(x)$ and $\csc(x)$

- Press $\boxed{\text{WINDOW}}$ to observe the settings for the trig viewing window.

The trigonometric viewing window gives a good picture of the graphs of $\sin(x)$ and $\csc(x)$. Keep in mind, however, that it is not the best window for viewing all trigonometric functions. For example, it would not be a good viewing window for the function $y = 5\cos(10x)$. For this function, you would need to adjust the window settings in order to display the key features of its graph.

Approximating Instantaneous Rates of Change

You can approximate the instantaneous rate of change of a function $f(x)$ at $x = a$ by evaluating the average rate of change of the function over a small interval containing a. For example, you might choose a symmetric interval about a of the form $(a - h, a + h)$ for some small value of h. Using this interval, your approximation of the instantaneous rate of change would be

$$\frac{f(a + h) - f(a - h)}{(a + h) - (a - h)} = \frac{f(a + h) - f(a - h)}{2h}.$$

Example: Use the previous formula with $h \doteq .01$ to approximate the instantaneous rate of change of $f(x) = \cos(x)$ at $x = \frac{\pi}{2}$ and $x = \pi$.

On the home screen, you'll enter the formula for the central difference quotient. The key strokes follow. Notice that both numerator and denominator are enclosed in parentheses. The formula is quite lengthy. So, don't press the enter key until you reach the end.

- On the home screen, enter the numerator for the first problem:
  Press (COS 2nd π ÷ 2 + . 0 1) .
  Press − COS 2nd π ÷ 2 − . 0 1)) .

- Press ÷ .

- Enter the denominator: press (2 × . 0 1)

- Press ENTER . You should get approximately − 1.0.

- Press 2nd ENTRY so that you can edit the previous problem.

- Edit the entry by changing $\pi/2$ to π. Then press ENTER . You should get 0. (Can you figure out why?)

Restricting the Domain of a Function

Only one-to-one functions have inverses. In order to define the inverse of a trigonometric function, you must restrict its domain so that the restricted trigonometric function is one-to-one. In the next example, you'll do that for the sine function.

Example: Graph $f(x) = \sin(x)$ on the restricted domain $-\frac{\pi}{2} \le x \le \frac{\pi}{2}$. Use settings for a [-2,2] × [-2,2] window.

Note: *Your calculator can only interpret one inequality at a time. Hence, it cannot interpret the condition $-\frac{\pi}{2} \le x \le \frac{\pi}{2}$ written in this form. Instead, you will need to enter the equivalent condition $-\frac{\pi}{2} \le x$ and $x \le \frac{\pi}{2}$.*

- Press Y= and clear any previously stored functions.

• Enter the sine function as Y1: press $\boxed{\text{SIN}}$ $\boxed{\text{X,T,}\theta\text{,}n}$ $\boxed{)}$. Don't press enter just yet.

• Add the restriction on the domain as follows:
Press $\boxed{\times}$ $\boxed{(}$ $\boxed{(-)}$ $\boxed{\text{2nd}}$ $\boxed{\pi}$ $\boxed{\div}$ $\boxed{2}$ $\boxed{\text{2nd}}$ $\boxed{\text{TEST}}$ $\boxed{6}$ $\boxed{\text{X,T,}\theta\text{,}n}$.
Press $\boxed{\text{2nd}}$ $\boxed{\text{TEST}}$, use the right arrow to highlight LOGIC, press $\boxed{1}$ for *and*.
Press $\boxed{\text{X,T,}\theta\text{,}n}$ $\boxed{\text{2nd}}$ $\boxed{\text{TEST}}$ $\boxed{6}$ $\boxed{\text{2nd}}$ $\boxed{\pi}$ $\boxed{\div}$ $\boxed{2}$ $\boxed{)}$ $\boxed{\text{ENTER}}$.

At the end of these key-strokes, your screen should match the one in Figure 38.

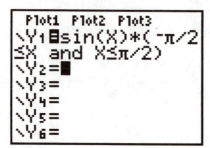

Figure 38. Restricting the Domain of $\sin(x)$

• Press $\boxed{\text{WINDOW}}$ and adjust the settings for a [-2,2] × [-2,2] window. Then press $\boxed{\text{GRAPH}}$. Notice that nearly vertical line segments connect the ends of the graph to the x-axis. To remove these line segments, press $\boxed{\text{MODE}}$ and change the settings from Connected to Dot. Then press $\boxed{\text{GRAPH}}$.

Notice that $f(x) = \sin(x)$ restricted to the domain $-\frac{\pi}{2} \le x \le \frac{\pi}{2}$ is one-to-one. Hence it has an inverse.

Do <u>not</u> clear this function from your calculator's memory until after you have completed the next example.

Example: Graph the restricted sine function from the previous example and its inverse using settings for a [-2,2] × [-2,2] window.
The restricted sine function, $f(x)$ from the previous example, should already be stored as Y1. Also, your calculator should be in Dot mode.

• Press $\boxed{\text{Y=}}$. Position the cursor opposite Y2.

• Press $\boxed{\text{2nd}}$ $\boxed{\text{SIN}^{-1}}$ $\boxed{\text{X,T,}\theta\text{,}n}$ $\boxed{)}$.

• Check that your window settings are for a [-2,2] × [-2,2] window. Then press $\boxed{\text{GRAPH}}$. Your screen should match Figure 39.

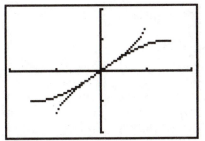

Figure 39. Graphs $\sin(x)$ and $\sin^{-1}(x)$

Finding Values of Inverse Trigonometric Functions

Example: Approximate the values of $\cos^{-1}(0.8)$ and $\cos^{-1}(1.8)$.
- To approximate $\cos^{-1}(0.8)$: press $\boxed{\text{2nd}}$ $\boxed{\cos^{-1}}$ $\boxed{.}$ $\boxed{8}$ $\boxed{)}$ $\boxed{\text{ENTER}}$. Your answer should be around .644.

- To compute $\cos^{-1}(1.8)$, press $\boxed{\text{2nd}}$ $\boxed{\text{ENTRY}}$, position the cursor over the decimal point, press $\boxed{\text{2nd}}$ $\boxed{\text{INS}}$ $\boxed{1}$ $\boxed{\text{ENTER}}$. This time you'll get an error message. That's because the input value of 1.8 is outside the domain of $\cos^{-1}(x)$. Press $\boxed{1}$ to quit.

Example: Find $\tan^{-1}(\frac{1}{2})$.
Press $\boxed{\text{2nd}}$ $\boxed{\text{TAN}^{-1}}$ $\boxed{1}$ $\boxed{\div}$ $\boxed{2}$ $\boxed{)}$ $\boxed{\text{ENTER}}$. You should get approximately .464.

Solving Equations Involving Trigonometric Functions

You can approximate solutions to equations involving trigonometric functions graphically.

Example: Approximate the solution to the equation $\cos(x) = 2\sin(x)$.
To approximate the solutions to this equation, first graph $y = \cos(x)$ and $y = 2\sin(x)$. Then approximate their points of intersection.

- Press $\boxed{\text{Y=}}$ and clear any previously stored functions. Then enter $\cos(x)$ as Y1 and $2\sin(x)$ as Y2.

- Press $\boxed{\text{ZOOM}}$ $\boxed{7}$. There are four intersection points, which correspond to four solutions, visible in this window. (See Figure 40.) If you widen the x-interval, you will see more intersection points, which correspond to additional solutions.

- Approximate the solution, the x-coordinate of the intersection point, that is closest to zero.
Press $\boxed{\text{2nd}}$ $\boxed{\text{CALC}}$ $\boxed{5}$ for intersect.
Press $\boxed{\text{ENTER}}$ to select the first curve and $\boxed{\text{ENTER}}$ again to select the second curve.
Press the right arrow key to move the cursor over the intersection point with x-value closest to zero. Your screen should be similar to the one in Figure 40. Then press $\boxed{\text{ENTER}}$. Read the x-value from the bottom of the screen. You should get approximately 0.464.

41

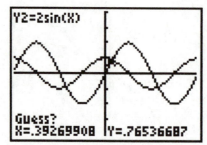

Figure 40. Guessing the Point of Intersection

• Next, repeat the steps outlined in the previous bullet to approximate the other solution that lies in the interval $0 < x < 2\pi$. You should get approximately 3.605.

Because the functions $\cos(x)$ and $2\sin(x)$ have period 2π, you can obtain other approximate solutions to $\cos(x) = 2\sin(x)$ by adding integer multiples of 2π to the two solutions that you've just found.

Continue directly to the next example to learn another method for solving this equation.

Example: Use your calculator's **solve** command to find the solutions to $\cos(x) = 2\sin(x)$ in the interval between 0 and 2π.

The syntax of the **solve** command is solve(*expression, variable, {lower bound, upper bound}*). This command yields a solution to the equation *expression* = 0. Hence, before using **solve**, you must replace the original equation with the equivalent equation $\cos(x) - 2\sin(x) = 0$. Here's how to find the solutions.

• Press $\boxed{\text{2nd}}$ $\boxed{\text{QUIT}}$ to return to the home screen.

• Approximate the solution between 0 and $\frac{\pi}{2}$ as follows:
   Press $\boxed{\text{2nd}}$ $\boxed{\text{CATALOG}}$.
   Press $\boxed{\text{S}}$ (same key as $\boxed{\text{LN}}$) to scroll down to the S's.
   Press the down arrow key to move the triangle opposite **solve(**. Then press $\boxed{\text{ENTER}}$.
   Press $\boxed{\cos}$ $\boxed{\text{X,T,}\theta\text{,}n}$ $\boxed{\text{)}}$ $\boxed{-}$ $\boxed{2}$ $\boxed{\sin}$ $\boxed{\text{X,T,}\theta\text{,}n}$ $\boxed{\text{)}}$ to enter the expression.
   Press $\boxed{\text{,}}$ $\boxed{\text{X,T,}\theta\text{,}n}$ $\boxed{\text{,}}$ to enter the variable.
   Press $\boxed{\text{2nd}}$ $\boxed{\{}$ $\boxed{0}$ $\boxed{\text{,}}$ $\boxed{\text{2nd}}$ $\boxed{\pi}$ $\boxed{\div}$ $\boxed{2}$ $\boxed{\text{2nd}}$ $\boxed{\}}$ $\boxed{\text{)}}$ to complete the command.
   Press $\boxed{\text{ENTER}}$ to execute the command. Your should get approximately .464.

Do <u>not</u> clear your screen until you have found the second solution.

• Approximate the solution between $\frac{\pi}{2}$ and 2π as follows:
   Press $\boxed{\text{2nd}}$ $\boxed{\text{ENTRY}}$.
   Change the lower and upper bounds to $\frac{\pi}{2}$ and 2π, respectively.
   Press $\boxed{\text{ENTER}}$. Your solution should match the one that you obtained graphically in the previous example.

Problems Inherent in the Technology
(Don't Believe Everything That You See!)

Your viewing screen consists of a grid of pixels. When a pixel is *on*, it shows up as a dark square dot on the screen. Graphs are formed by turning on a series of pixels. This method of producing graphs can sometimes produce misleading images.

Example: Graph $g(x) = \sin(x)$ over increasingly wide x-intervals.
- Press $\boxed{Y=}$ and clear any previously stored functions.

- Now, begin by entering $g(x)$ as Y1. Then press $\boxed{ZOOM}$ $\boxed{7}$ to graph $g(x)$ in the trigonometric viewing window.

- Turn off the tick marks for the x-axis: press $\boxed{WINDOW}$ and set Xscl = 0.

- Change the zoom factors: press $\boxed{ZOOM}$, use the right arrow key to highlight Memory, then press $\boxed{4}$ for SetFactors.

- Set XFact = 10 and YFact = 1 and then press $\boxed{GRAPH}$ to return to the graph.

Get ready to have some fun. When $\sin(x)$ is graphed in the trig viewing window, how many complete sine waves do you see? Each time you increase the width of the x-interval by a factor of 10, you should see 10 times as many cycles of the sine wave.

- Press $\boxed{ZOOM}$ $\boxed{3}$ for Zoom Out. Then press $\boxed{ENTER}$. Do you see 10 times as many cycles? Press $\boxed{ENTER}$ to zoom out again. Do you see 10 times as many cycles as in the previous graph?

    The graph produced by the second zoom out shows fewer cycles than the graph produced by the first zoom out. In producing the graphs, your calculator does not have enough pixels to capture all the oscillations that are part of the actual graph. In this case, the small subset of points from the actual graph that your calculator chooses to represent with darkened pixels presents a very misleading picture of the features of the actual graph.

- Press $\boxed{ENTER}$ several more times to view other interesting graphs, all of which give misleading pictures of the function's key graphical features.

CHAPTER 7

Finding Values of Trigonometric Functions

With the TI-83Plus you can solve problems in right-triangle trigonometry. To compute the sine, cosine, or tangent of an angle measured in degrees, you will first need to change the mode setting to Degree. Press MODE , use the arrow keys to position the cursor on Degree, and press ENTER to save the setting. Press 2nd QUIT to return to the home screen.

Example: Compute sin(60°) and sin(29°).
- Press SIN 60) ENTER . You should get approximately 0.866.

- Edit the previous command to compute sin(29°): press 2nd ENTRY , change 60 to 29, and press ENTER .

Example: Compute tan(90°).

Press TAN 9 0) ENTER . Your calculator will respond with the error message: ERR: DOMAIN because the tangent function is undefined at 90°. Press 1 to quit.

Finding Values of Inverse Trigonometric Functions

Example: Approximate the value of sin⁻¹(.3).
Press 2nd SIN⁻¹ . 3) ENTER . You should get approximately 17.5°. (Because you are in degree mode, the output will be in degrees.)

Example: Find the approximate value of tan⁻¹(10).
Press 2nd TAN⁻¹ 1 0) ENTER . You should get approximately 84.3°.

CHAPTER 8

Working With Parametric Equations

Before you can use your calculator to graph parametric equations, you'll need to change your calculator from function mode to parametric mode. Press MODE and adjust your mode settings to match those in Figure 41. Then press 2nd QUIT to return to the home screen.

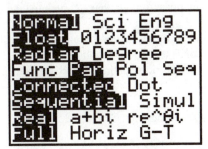

Figure 41. Mode Settings

Example: Graph the parametric equations $x(t) = 10 - 3t$ and $y(t) = 1 + 4t$ in the standard viewing window.

- Press Y=. Notice in parametric mode, you enter pairs of formulas, one for $x(t)$, and the other for $y(t)$. If you have any parametric equations stored in memory, erase them by positioning the cursor on each equation and pressing CLEAR.

- Enter the formula for $x(t)$ as X₁ₜ: press 1 0 − 3 X,T,θ,n ENTER.

- Enter the formula for $y(t)$ as Y₁ₜ: press 1 + 4 X,T,θ,n ENTER.

- Press ZOOM 6 to graph this set of parametric equations in the standard viewing window.

- Press WINDOW to observe the settings for the standard window. Your screen should match the one in Figure 42. (If Tmax = 360 instead of approximately 6.28, change your mode settings to Radian.)

```
WINDOW
 Tmin=0
 Tmax=6.2831853…
 Tstep=.1308996…
 Xmin=-10
 Xmax=10
 Xscl=1
↓Ymin=-10
```

Figure 42. Standard Window Settings

Notice that the settings for the standard viewing window include settings for t: Tmin = 0, Tmax $\approx 2\pi$, and Tstep $\approx \frac{\pi}{24}$. The settings for the x-interval and y-interval are the same as in function mode.

• Press $\boxed{\text{GRAPH}}$ to return to the graph and then press $\boxed{\text{TRACE}}$. Notice that the trace begins at the point corresponding to t = Tmin, or in this case $t = 0$. Press the right arrow to move along the graph in t-increments of approximately 0.13.

Do not clear these parametric functions until you have completed the next example.

Example: Graph the position of a dot as it moves along the path $x(t) = 10 - 3t$ and $y(t) = 1 + 4t$ at one-second intervals from time $t = 0$ seconds to $t = 10$ seconds.

You should already have the parametric equations stored in your calculator from your work on the previous example.

• Press $\boxed{\text{MODE}}$ and change the setting from Connected to Dot.

• Press $\boxed{\text{WINDOW}}$. Adjust the parameter settings for t: Tmin = 0, Tmax = 10, Tstep = 1.

• Press $\boxed{\text{ZOOM}}$, use the down arrow key to highlight ZoomFit and then press $\boxed{\text{ENTER}}$. You should see a group of isolated dots that fall on a line similar to those shown in Figure 43.)

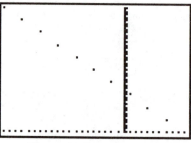

Figure 43. Dots On a Line

• Press $\boxed{\text{TRACE}}$. Then hold down the right arrow key. Watch the cursor move from one dot to the next.

Return your calculator to Connected mode before beginning the next topic.

Using Square Scaling in Parametric Mode

Example: Graph the circle described by the set of parametric equations $x(t) = 5\cos(t)$, $y(t) = 5\sin(t)$ first in the standard viewing window and then in a square-scaling window.
• Press $\boxed{\text{Y=}}$ and clear any stored equations. Then enter the formula for $x(t)$ as X_{1T} and the formula for $y(t)$ as Y_{1T}.

- Press $\boxed{\text{ZOOM}}$ $\boxed{6}$. Your graph should look more elliptical than circular.

- Press $\boxed{\text{ZOOM}}$ $\boxed{5}$. Your graph should now look like a circle.

Combining Two Sets of Parametric Equations

In Lab 8B, Bézier Curves, you are asked to form a new set of parametric equations from a combination of two other sets of parametric equations.

Example: Suppose that you have two sets of parametric equations

$$\text{S1: } x_1 = 2t + 1 \qquad \text{S}_2\text{: } x_2 = t - 5$$
$$y_1 = -3t + 5 \qquad\qquad y_2 = 4t - 3$$

and that you want to graph the combination $(1 - t)\text{S}_1 + t\text{S}_2$ over the interval $0 \leq t \leq 1$.

You'll tackle this problem in three steps. In Step 1, you'll enter the equations for S_1 and S_2. In Step 2, you'll form the combination. In Step 3, you'll graph the combination.

Step 1: Enter the parametric equations S_1 and S_2.
- Be sure that your calculator is in parametric mode. Press $\boxed{\text{Y=}}$ and clear any stored functions.

- Enter the equations for S_1 as X_{1T} and Y_{1T} and the equations for S_2 as X_{2T} and Y_{2T}.

Step 2: Enter the x- and y-equations for the combination $(1 - t)\text{S}_1 + t\text{S}_2$:

$$x_3 = (1 - t)x_1 + tx_2$$
$$y_3 = (1 - t)y_1 + ty_2$$

- The cursor should be opposite X_{3T}. Enter the equation for x_3:
  Press $\boxed{(}$ $\boxed{1}$ $\boxed{-}$ $\boxed{\text{X,T},\theta,n}$ $\boxed{)}$ $\boxed{\times}$.
  Press $\boxed{\text{VARS}}$, use the right arrow key to highlight Y-VARS, press $\boxed{2}$ for Parametric, and
  then press $\boxed{1}$ $\boxed{(}$ $\boxed{\text{X,T},\theta,n}$ $\boxed{)}$ for $x_1(t)$.
  Press $\boxed{+}$ $\boxed{\text{X,T},\theta,n}$ $\boxed{\times}$.
  Press $\boxed{\text{VARS}}$, use the right arrow key to highlight Y-VARS, press $\boxed{2}$ for Parametric, and
  then press $\boxed{3}$ $\boxed{(}$ $\boxed{\text{X,T},\theta,n}$ $\boxed{)}$ for $x_2(t)$.
  Press $\boxed{\text{ENTER}}$.

- The cursor should now be opposite Y_{3T}. Enter the equation for y_3.
  Press $\boxed{(}$ $\boxed{1}$ $\boxed{-}$ $\boxed{\text{X,T},\theta,n}$ $\boxed{)}$ $\boxed{\times}$
  Press $\boxed{\text{VARS}}$, use the right arrow key to highlight Y-VARS, press $\boxed{2}$ for Parametric, and
  then press $\boxed{2}$ $\boxed{(}$ $\boxed{\text{X,T},\theta,n}$ $\boxed{)}$ for $y_1(t)$.
  Press $\boxed{+}$ $\boxed{\text{X,T},\theta,n}$ $\boxed{\times}$.
  Press $\boxed{\text{VARS}}$, use the right arrow key to highlight Y-VARS, press $\boxed{2}$ for Parametric, and
  then press $\boxed{4}$ $\boxed{(}$ $\boxed{\text{X,T},\theta,n}$ $\boxed{)}$ for $y_2(t)$.

Step 3: Graph the combination $(1 - t)S_1 + tS_2$. Here's how:

- Unselect (turn off) parametric equations $xt1$, $yt1$, $xt2$, and $yt2$. To do this, move the cursor over the equals sign opposite X_{1T}. Press ENTER . Notice that both X_{1T} and Y_{1T} have been turned off. Repeat this process to turn off X_{2T} and Y_{2T}.

- Press WINDOW . Set Tmin = 0, Tmax = 1, and Tstep = 0.1. Adjust the remainder of the settings for a $[-5,2] \times [-4,6]$ window.

- Press GRAPH to graph the combination. Your graph should match the one in Figure 44.

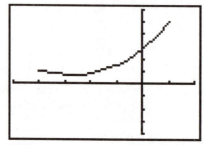

Figure 44. Graphing a Combination

You've reached the end of this guide. Your calculator has many more features in addition to what you've learned from this guide. Consult the manual that came with your calculator to learn more about your calculator's capabilities.

Graphing Calculator Guidebook
for the TI-85/86
To Accompany
Precalculus: Concepts in Context, 2e

This guide provides background on the TI-85/86 graphing calculator that will be useful for *Precalculus: Concepts in Context, 2e*. It consists of a basic tutorial followed by additional instructions relevant to each chapter of your text. Consult your calculator manual for additional calculator features.

BASIC TUTORIAL

0. *The Keyboard*

Most of the keys on the TI-85/86 access more than one function.
- To access a function written in yellow lettering, press the | 2nd | key.
  Example: | 2nd || QUIT | (same key as | EXIT |).
- To access a function written in blue, press the | ALPHA | key.
  Example: | ALPHA || = | (same key as | STO ▷ |).

Some keys, such as | GRAPH |, bring up menus. The menu keys | F1 || F2 || F3 || F4 || F5 | provide access to menu items. These keys are located directly below the calculator screen.

You should familiarize yourself with the arrow keys, the keys that control the cursor. These keys are located on the right-hand side of the calculator (below | F4 | and | F5 |) and point out in four directions as shown in Figure 1.

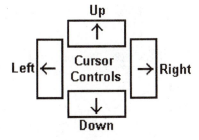

Figure 1. The Arrow Keys

Finally, take a moment to look at the last column of keys on the right side of the calculator. Here you will find the operation keys for addition, subtraction, multiplication, division, and powers. You will also find the | CLEAR | and | ENTER | keys in this column.

TI-85/86

1. *On, Off, and Contrast*

Turn the calculator on by pressing $\boxed{\text{ON}}$.

You may need to adjust the contrast. Press the $\boxed{\text{2nd}}$ key followed by holding down the up arrow key to darken or the down arrow key to lighten.

To turn your calculator off, press $\boxed{\text{2nd}}$ $\boxed{\text{OFF}}$. If you forget, the calculator will automatically turn off after a period of non-use.

2. *Basic Calculations*

The screen that displays your calculations is called the home screen. If your screen shows a menu or a graph when it is turned off, in some cases, it will return to this menu or graph when you turn it back on. Press $\boxed{\text{EXIT}}$ one or more times, or press $\boxed{\text{2nd}}$ $\boxed{\text{QUIT}}$, to return to the home screen.

At times, you'll want to start with a clear home screen. To remove previous calculations, press $\boxed{\text{CLEAR}}$. You do not need to clear the home screen after each computation. If not cleared, the home screen will keep a history of several lines of your work.

Work through the following examples to learn about basic computations and techniques on the TI-85/86.

Example: Compute 3×4.
    After pressing $\boxed{3}$ $\boxed{\times}$ $\boxed{4}$, press $\boxed{\text{ENTER}}$. Note that the original problem, written as $3 * 4$, remains on the left side of the screen and the answer appears on the right.

Example: Compute $3 + 2 \times 6$ and then $(3 + 2) \times 6$.
    After you complete the first calculation, you'll use your calculator's editing features to save keystrokes when entering the second calculation.

• First, press $\boxed{3}$ $\boxed{+}$ $\boxed{2}$ $\boxed{\times}$ $\boxed{6}$ $\boxed{\text{ENTER}}$. Do not press $\boxed{\text{CLEAR}}$.

• Complete the second calculation as follows:
    Press $\boxed{\text{2nd}}$ $\boxed{\text{ENTRY}}$ (same key as $\boxed{\text{ENTER}}$) to rewrite the original calculation.
    Press the left arrow key to move the cursor (a blinking rectangle) over the 3.
    Press $\boxed{\text{2nd}}$ $\boxed{\text{INS}}$ (same key as $\boxed{\text{DEL}}$). The cursor turns into a blinking underline. Then press $\boxed{(}$ to insert the parenthesis to the left of 3.
    Press the right arrow key to move the cursor over the *. Press $\boxed{\text{2nd}}$ $\boxed{\text{INS}}$ $\boxed{)}$.
    Press $\boxed{\text{ENTER}}$. You should get 30.

Example: Compute 8^3.
    Press $\boxed{8}$ followed by $\boxed{\wedge}$ $\boxed{3}$ $\boxed{\text{ENTER}}$.

Example: Compute $\sqrt{16}$.
    Press ⌈2nd⌉ ⌈[$\sqrt{}$]⌉ (same key as ⌈x^2⌉) followed by ⌈1⌉ ⌈6⌉ ⌈ENTER⌉.

Example: Compute $\sqrt{-16}$.
    Warning! *The TI-85/86 has two minus keys,* ⌈ $-$ ⌉ *and* ⌈(-)⌉, *to differentiate between the operation of subtraction (such as $3 - 2 = 1$) and the opposite of the positive number 16, namely -16.*

    Press ⌈2nd⌉ ⌈[$\sqrt{}$]⌉ followed by ⌈(-)⌉ (the key to the left of ⌈ENTER⌉) ⌈1⌉ ⌈6⌉ ⌈ENTER⌉.

    Note: *Many calculators would report an error message when you tried to compute $\sqrt{-16}$ because there is no real number whose square is -16. However, your TI-85/86 calculator does the computation using complex numbers. It responds (0,4), which represents the number $0 + 4\sqrt{-1}$ or $0 + 4i$.*

Example: Find a decimal approximation of $\sqrt{5}$.
    Press ⌈2nd⌉ ⌈ $\sqrt{}$ ⌉ ⌈5⌉ ⌈ENTER⌉. In this case, your calculator provides a decimal approximation of $\sqrt{5}$, which is around 2.236. This is not an exact answer (even though the answer on your screen may be precise to 10 decimal places!).

Example: Compute $\sqrt[5]{32}$.
    • Press ⌈5⌉ for the fifth root.

    • Press ⌈2nd⌉ ⌈MATH⌉ (same key as times, ⌈ $\times$ ⌉) to access the math menu, then press ⌈F5⌉ to choose MISC. The first five choices of the MATH/MISC menu are displayed on the bottom line of the menu. (Your screen should be similar to the one in Figure 2.)

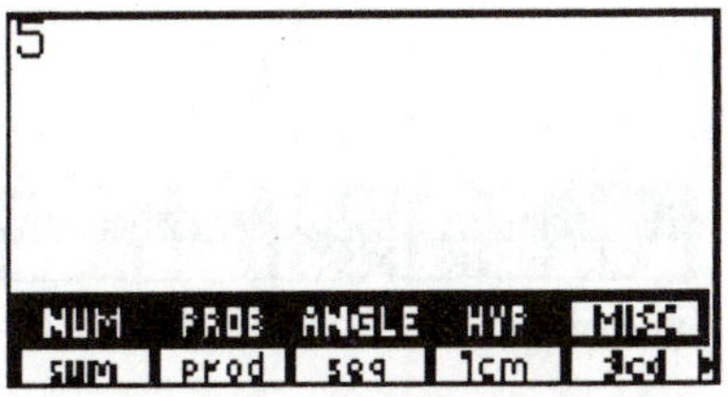

Figure 2. The MATH/MISC Menu

    • Press ⌈MORE⌉ to reveal more options. The root operation, $\sqrt[x]{}$, will now appear above the ⌈F4⌉ key. Press ⌈F4⌉ to select $\sqrt[x]{}$.

    • Press ⌈3⌉ ⌈2⌉ ⌈ENTER⌉. Did you get 2?

    • Press ⌈EXIT⌉ twice, once to exit the MISC submenu, and the second time to exit the MATH menu.

3. Correcting an Error and Editing

If you make an error in calculation, you may be able to correct the error without reentering the problem. You can also use the editing feature to modify previous calculations. Check out the following two examples.

Example: Make a deliberate error by entering 3 + + 2 and then correct it.
Press ⎡3⎤ ⎡+⎤ ⎡+⎤ ⎡2⎤ ⎡ENTER⎤. To correct the error, press ⎡F1⎤ for GOTO. The cursor will direct you to the error. Erase one of the plus signs by pressing ⎡DEL⎤, and then press ⎡ENTER⎤. The correct answer to 3 + 2 will appear.

Example: Enter the problem (3 + 2 × 6)/5. Then change it to (3 + 2 × 6)/7.
- Enter the first problem: press ⎡(⎤ ⎡3⎤ ⎡+⎤ ⎡2⎤ ⎡×⎤ ⎡6⎤ ⎡)⎤ ⎡÷⎤ ⎡5⎤ ⎡ENTER⎤.

- Press ⎡2nd⎤ ⎡ENTRY⎤ (same key as ⎡ENTER⎤). To rewrite the previous problem on your screen.

- Press the left arrow key to move the cursor (a solid rectangle) over 5, and then press ⎡7⎤ ⎡ENTER⎤.
  (Because the cursor was a solid rectangle, you overwrote the 5 when you pressed ⎡7⎤.)

4. Resetting the Memory and Changing the Mode Settings

After experimenting with settings, you may want to return your calculator's settings to the factory settings. Here's how.
- Press ⎡2nd⎤ ⎡MEM⎤ (on the TI-86, same key as ⎡3⎤; on the TI-85, same key as ⎡+⎤) ⎡F3⎤ to access the MEM/RESET menu. Your screen should be similar to Figure 3.

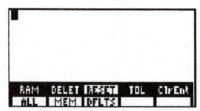

Figure 3. MEM/RESET Menu on TI-86

MEM clears the memory (programs, data, functions, graphs) but leaves the mode settings alone. DFLTS leaves the memory alone but returns the mode settings to their defaults. ALL resets both memory and defaults.

- Press ⎡F3⎤ for DFLTS and then ⎡F4⎤ for YES. The message "Defaults set" should appear on your screen. If you can't read this message, darken the contrast. (Instructions for adjusting the contrast appear in Topic 1, page 50.)

- Press ⎡CLEAR⎤ to clear the screen.

Example: Compute 250×157. Display the results in scientific notation.

First, you'll change the mode settings from Normal to Sci (for scientific notation). Then you'll do the computation.

- Press 2nd MODE (same key as MORE). The cursor should be blinking on Normal. If you have reset your calculator, the mode settings will match the default settings shown in Figure 4.

Figure 4. Mode Menu

- Use the right arrow key to move the cursor over Sci and press ENTER to change the setting.

- Press 2nd QUIT (same key as EXIT) to return to the home screen.

- Press 2 5 0 × 1 5 7 ENTER . Your answer should be 3.925 E4, shorthand for 3.925×10^4.

- Press 2nd MODE and return the setting to Normal. Press 2nd QUIT to return to the home screen.

5. Graphing

First, check your mode settings: press 2nd MODE . If necessary, adjust the mode settings to match the default settings in Figure 4. Press 2nd QUIT to return to the home screen.

Example: Graph $y = x$, $y = x^2$, and $y = x^3$ in the standard viewing window.

You'll tackle this problem in two steps. In step 1, you'll enter the function. In step 2, you'll graph it and then look at the window settings.

Step 1: Enter the three functions as follows.

- Press GRAPH F1 to view the GRAPH/$y(x) =$ screen.

- If you have functions stored in your calculator, use the up or down arrow keys to position the cursor on the same line as the function. Then press CLEAR to erase the function.

- Enter the first function to the right of $y1$ by pressing F1 or x-VAR for x.

- Press ENTER to move the cursor to the right of $y2$. Enter the second and third functions to the right of $y2$ and $y3$, respectively. (Remember to press ^ to obtain the powers.)

53

• Press ☐EXIT☐ to exit the $y(x) =$ submenu and return to the GRAPH menu.

Step 2: Set the standard viewing window and graph the functions. Then check the window settings. Here's how.

• Press ☐F3☐ to display the GRAPH/ZOOM menu. Then press ☐F4☐ to select ZSTD. The graphs of all three functions should appear on your screen. To remove the menu bar from the bottom of the screen, press ☐CLEAR☐.

• Press ☐GRAPH☐ to return to the GRAPH menu. To access the window settings, press ☐F2☐. You are now in the GRAPH/WIND submenu on the TI-86 or the GRAPH/RANGE submenu on the TI-85. Your screen should be similar to the screen in Figure 5.

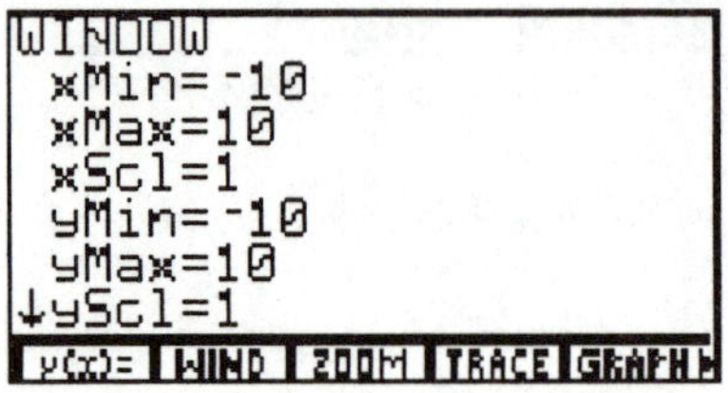

```
WINDOW
 xMin=-10
 xMax=10
 xScl=1
 yMin=-10
 yMax=10
↓yScl=1
 y(x)= WIND ZOOM TRACE GRAPH▸
```
Figure 5. Accessing the Window Settings

• To exit this menu and return to the home screen, press ☐EXIT☐ or ☐2nd☐ ☐QUIT☐.

Do not erase the functions you stored as $y1$, $y2$ and $y3$ until you have completed this tutorial.

Example: Graph $y = x$, $y = x^2$, and $y = x^3$, the functions from the previous example, using the window settings shown in Figure 6.

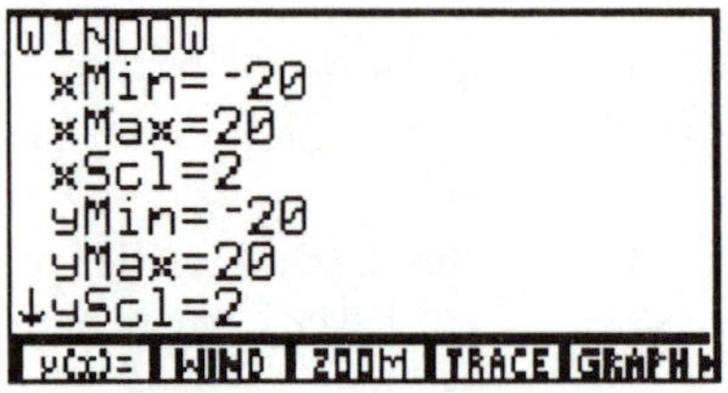

```
WINDOW
 xMin=-20
 xMax=20
 xScl=2
 yMin=-20
 yMax=20
↓yScl=2
 y(x)= WIND ZOOM TRACE GRAPH▸
```
Figure 6. New Window Settings

• Press ☐GRAPH☐ ☐F2☐ to access the window settings. Change the first setting to -20 (*remember to press* ☐(-)☐ *when you enter* -20) and press ☐ENTER☐. Change the remaining settings to match those in Figure 6.

• Press ☐F5☐ to view the graphs in the new window.

Continue directly to the next example.

Example: Zoom in twice on the graph from the previous example to get a magnified view around a point of intersection. Then use Trace to estimate the coordinates of this point.

- From the GRAPH menu, press $\boxed{\text{F3}}$ for ZOOM and then $\boxed{\text{F2}}$ for ZIN, which is short for zoom in. Notice that the center of this first zoom (look for the blinking pixel) will be (0,0). Press $\boxed{\text{ENTER}}$ to zoom in on the origin. Your graph should be similar to the one in Figure 7.

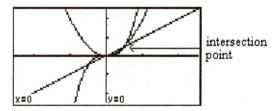

Figure 7. Intersection Point

- Notice the point of intersection to the right of (0,0). (See Figure 7.) Next, you'll zoom in using a center close to this point of intersection. Press the right and up arrow keys to move the cursor (a plus sign with a blinking center pixel) close to this point of intersection. Then press $\boxed{\text{ENTER}}$. The intersection point should be clearly visible near the center of the screen.

- Press $\boxed{\text{GRAPH}}$ $\boxed{\text{F4}}$ for TRACE. Trace along the first curve by pressing the left and right arrow keys. You should see the cursor (a square with a blinking $\times$) move along the curve. Press the up or down arrow keys to jump from one curve to another.

- Move the cursor directly on top of the intersection point that lies to the right of the origin. Read the approximate values of the x- and y-coordinates from the bottom of the screen. (The coordinates should be close to (1,1).)

Example: Take a closer look at the graph of $y = x$ without the graphs of the other functions. Then view the graph in a window that uses square scaling.

You'll tackle this problem in three steps. In Step 1, you'll turn off the functions stored as $y2$ and $y3$. In Step 2, you'll set up the window and view the graph. In Step 3, you'll clear the stored functions.

Step 1: Remove the graphs of $y = x^2$ and $y = x^3$ from the viewing screen without erasing these functions.

- Press $\boxed{\text{GRAPH}}$ $\boxed{\text{F1}}$ to display the stored functions.

- To unselect $y2$, use the up or down arrow key to move the cursor to the line containing $y2$'s formula. Press $\boxed{\text{F5}}$ for SELCT, which is short for select, to remove the highlighting over the equals sign.

- Next, move the cursor to the line containing $y3$'s formula and press $\boxed{\text{F5}}$.

Step 2: View the graph of $y = x$ in the standard viewing window and then switch to square scaling.

- Press $\boxed{\text{EXIT}}$ to return to the GRAPH menu. Then press $\boxed{\text{F3}}$ $\boxed{\text{F4}}$ to view the graph of $y = x$ in the standard viewing window.

The line should make a 45° angle with the x-axis, but the apparent angle on this screen is less than 45°. (Note that the tick marks on the y-axis are much closer together than on the x-axis.)

- You should still be in the GRAPH menu. Press $\boxed{\text{F3}}$ $\boxed{\text{MORE}}$ $\boxed{\text{F2}}$ for ZSQR. Press $\boxed{\text{CLEAR}}$ to remove the menu bars.

Now you should see a true 45° angle. (Observe that the tick marks on the x- and y-axes are equally spaced.)

Note: *You can turn the functions for y2 and y3 back on by repeating Step 1. This time the process will restore the highlighting on the equals signs in the formulas for y2 and y3.*

Step 3: Clear the stored functions.
- Press $\boxed{\text{GRAPH}}$ $\boxed{\text{F1}}$.

- The cursor should be on the same line as $y1$. Press $\boxed{\text{CLEAR}}$. Then use the down arrow to position the cursor on the same line as $y2$ and press $\boxed{\text{CLEAR}}$. Repeat for $y3$.

That's it! You have completed the tutorial. Now practice and experiment on your own with the calculator until you feel comfortable with these basic operations. The remainder of this guide will introduce new techniques and occasionally review techniques as they are needed, chapter by chapter, for your work in *Precalculus: Concepts in Context, 2e.*

Chapter-by Chapter Guide

CHAPTER 1

Returning to and Clearing the Home Screen

To exit a menu or submenu and return to the home screen, press ⟨EXIT⟩ one or more times. To return to the home screen from a graph press ⟨2nd⟩ ⟨QUIT⟩ (same key as ⟨MODE⟩). On a TI-86, you can also exit menus by pressing ⟨2nd⟩ ⟨QUIT⟩. To clear the home screen, press ⟨CLEAR⟩ once or twice.

Plotting Points

You can use your calculator to plot the Fahrenheit-Celsius data from Lab 1A. Then graph your guess for the formula that relates degrees Fahrenheit to degrees Celsius. This will allow you to check how closely the function specified by your formula follows the pattern of the data.

Example: Plot the data in Table 1 and then overlay the graph of $y = 18x + 85$.

You'll work on this problem in six steps. In Step 1 and Step 2, you'll clear any previously stored functions and data. In Step 3, you'll enter the data from Table 1. In step 4, you'll plot the data and in Step 5, you'll add a line to your plot. Finally, in Step 6, you'll turn off the data plot.

Sample Data	
x	y
-2	40
-1	60
1	100
3	140

Table 1. Sample Data

Step 1: Clear any stored functions.

- Press ⟨GRAPH⟩ ⟨F1⟩. Clear any stored functions by positioning the cursor on a line with a stored function and pressing ⟨CLEAR⟩.

- Return to the home screen by pressing ⟨2nd⟩ ⟨QUIT⟩ or ⟨EXIT⟩ twice.

Step 2: Clear any stored data.

The TI-85/86 has built-in list names for the x-variable list (xStat) and the y-variable list (yStat). First, check to see if any data have been stored in these lists, and, if so, clear the lists.

On the TI-86:

- Press ⬚2nd ⬚STAT (same key as ⬚+) followed by ⬚F2 for EDIT. You should see three lists: xStat, yStat, and fStat. If no data appear in the xStat and yStat lists, you can proceed directly to Step 3. Otherwise, complete Step 2.

- If data are stored in these lists, use the arrow keys to highlight xStat. Then press ⬚CLEAR followed by ⬚ENTER . Now do the same for the yStat and fStat lists.

On the TI-85:

- Press ⬚STAT ⬚F2 for EDIT. Press ⬚ENTER twice to get to the xStat and yStat lists. If no data have been stored, the cursor will be blinking to the left of a blank "$x1 =$". The next line will read: $y1 = 1$. In this case, proceed to Step 3. Otherwise complete Step 2.

- If data are stored in these lists, erase the data by pressing ⬚F5 for CLRxy.

Step 3: Enter the data from Table 1.

After completing Step 2, you should be in the xStat and yStat lists. Your screen should resemble either Figure 8 or Figure 9.

Figure 8. Stat Lists on the TI-85

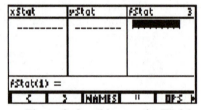

Figure 9. Stat Lists on the TI-86

On the TI-86:

- If necessary, use your arrow keys to highlight the cell directly beneath xStat (as shown in Figure 10). Enter the x-data: press ⬚(-) ⬚2 ⬚ENTER for the first entry. Then enter the remaining three values for x.

- Press the right arrow key to highlight the cell directly beneath yStat. Then enter the y-data.

On the TI-85:

- Enter the data one x-y pair at a time. (Initially, the cursor should be blinking opposite $x1$. Press ⬚(-) ⬚2 ⬚ENTER ⬚4 ⬚0 ⬚ENTER for the first ordered pair.

- Press ⬚(-) ⬚1 ⬚ENTER ⬚6 ⬚0 ⬚ENTER for the second; press ⬚1 ⬚ENTER ⬚1 ⬚0 ⬚0 ⬚ENTER for the third; and press ⬚3 ⬚ENTER ⬚1 ⬚4 ⬚0 ⬚ENTER for the fourth.

- Press ⬚2nd ⬚QUIT to return to the home screen.

Step 4: Plot the sample data.

On the TI-86:

- Press $\boxed{\text{2nd}}$ $\boxed{\text{STAT}}$ and then $\boxed{\text{F3}}$ to select PLOT.

- Press $\boxed{\text{F1}}$ to choose PLOT 1. To turn the plot on, press $\boxed{\text{ENTER}}$. Your screen should match the one in Figure 10. If the Type does not match the one in Figure 10, press the down arrow key and then $\boxed{\text{F1}}$ for SCAT. You can change the lists and the symbol for the mark in a similar fashion.

Figure 10. Screen for Plot 1

- Next, adjust the window settings: press $\boxed{\text{GRAPH}}$ $\boxed{\text{F3}}$ for ZOOM and then press $\boxed{\text{MORE}}$ $\boxed{\text{F5}}$ for ZDATA. Press $\boxed{\text{CLEAR}}$ to remove the menu bar from the bottom of your screen.

- If your y-axis appears thick, you may want to change the setting for yScl. Press $\boxed{\text{GRAPH}}$ $\boxed{\text{F2}}$. Use the down arrow to move the cursor opposite yScl. Enter yScl = 10. Then press $\boxed{\text{F5}}$ $\boxed{\text{CLEAR}}$. Your graph should resemble Figure 11.

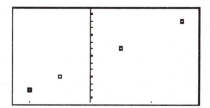
Figure 11. Scatterplot of Data From Table 1

On the TI-85:

- First, you'll need to adjust the window settings: press $\boxed{\text{GRAPH}}$ $\boxed{\text{F2}}$ to access the GRAPH/RANGE screen. Choose a value for xMin that is smaller than the x-coordinates in Table 1 and a value for xMax that is larger than the x-coordinates. Similarly, select appropriate settings for yMin and yMax. Decide on the spacing of the tick marks and set xScl and yScl.

- Next, you'll plot the data in the window that you have selected. Press $\boxed{\text{STAT}}$ $\boxed{\text{F3}}$ to select DRAW and then $\boxed{\text{F2}}$ for SCAT (short for scatterplot). Press $\boxed{\text{CLEAR}}$ to clear the menu bar from the scatterplot. If you have chosen appropriate settings for xMin, xMax,

yMin, and yMax, you should see a plot of four points similar to Figure 11. Note that the TI-85 marks a point with a single pixel, which can be difficult to see.

Step 5: Add the graph of $y = 18x + 85$ to the plot of the sample data.
- Press ⬚GRAPH ⬚F1 to enter the GRAPH/$y(x) =$ menu. Enter $18x + 85$ as $y1$.

- Press ⬚EXIT to return to the GRAPH menu and then press ⬚F5 ⬚CLEAR to view the graph. If you are using a TI-86, the graph of the line will be added to the scatterplot from Step 4 and you can skip directly to Step 6.

  If you are using a TI-85, you see the graph of $y = 18x + 85$ but the scatterplot has disappeared from your screen. To overlay the scatterplot, press ⬚STAT ⬚F3 ⬚F2 ⬚CLEAR. The plotted points should lie close to the line.

Step 6. Turn off the data plot.
On a TI-86, press ⬚2nd ⬚STAT ⬚F3 ⬚F5 ⬚ENTER. Your calculator will respond "Done."

On a TI-85, the scatterplot will disappear after you have changed your function or window settings.

Warning! *If you are using a TI-86 and you fail to turn off Plot 1, this plot will appear superimposed on your next graph. Worse, if you subsequently erase the data in xStat and yStat without turning off this plot, you will get an error message each time you press* ⬚GRAPH *because your calculator will still be trying to create STAT PLOT 1.*

Making a Table of Values from a Formula

The TI-86 has a table feature, which provides numeric information about a function. Unfortunately, the TI-85 does not have a built in table feature; however, you can download a table program from the TI website. Directions for downloading a table program for the TI-85 follow the example.

In the next example, you'll start with a column of values for the independent variable, x, and calculate the corresponding values for the dependent variable, y. These directions are for the TI-86 but can be adapted for use with table programs for the TI-85.

Example: Use your TI-86 to complete Table 2 for the function $y = 2x - 30$.

x	y
2.0	
6.0	
10.0	
14.0	
18.0	
22.0	
26.0	

Table 2. A Table of Function Values

Plan of action: First, you'll enter the function into the calculator. Then you'll set up the values for the independent variable, x. Notice that the minimum x-value in the table is 2.0 and that consecutive x-values are separated by increments of 4.0. Using the TABLE menu, you'll be able to generate the values for the x- and y-columns.

• Press $\boxed{\text{GRAPH}}$ $\boxed{\text{F1}}$. Clear any previously stored functions and then enter the function $y = 2x - 30$ as $y1$.

• Press $\boxed{\text{TABLE}}$ $\boxed{\text{F2}}$ for TBLST to access the TABLE SETUP screen.

• Change the settings to match those in Figure 12.

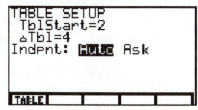

Figure 12. Setting Up a Table

• To view the table, press $\boxed{\text{F1}}$ for TABLE.

Using the up and down arrows, you can scroll up or down in the table.

For example, use the down arrow key to find the value of $y1$ that corresponds to an x-value of 62. (You should get 94.)

Now, use the up arrow key to find two x-values in the table for which the dependent variable change signs. (You should get x-values of 14 and 18.)

• Press $\boxed{\text{2nd}}$ $\boxed{\text{QUIT}}$ to exit the table and return to the home screen.

Directions for Downloading a Table Program for the TI-85: If you have a graph link cable, you can download a table program from Texas Instruments' program archives. In order to transfer the program from a computer to your calculator, you will need to install TI-85 Graph Link software as well. Here's information on downloading a table program.
   • The url for the Texas Instruments website is www.ti.com.
   • In the Keyword search box, enter TI-85 and click Go.
   • Click Downloads--TI-85.
   • Click TI-85 Program Archive.
   • Open the Math folder. There are several table programs to choose from. The .txt files contain documentation for each of the programs.
   • Download one of the table programs.
   • Using the graph link software, transfer the table program from your computer to your calculator. (If you don't have a graph link cable, use the graph link software to open the table program file so you can read the code. Then key in the program manually.)

Computing the Value of a Function

Example: Find the value of the function $f(x) = 3x^2 + 6x - 7$ for $x = 5$ and $x = 8$. In other words, find the values of $f(5)$ and $f(8)$.

- Press $\boxed{\text{GRAPH}}$ $\boxed{\text{F1}}$ and erase any previously stored functions. Enter the formula for $f(x)$ as $y1$. Then press $\boxed{\text{2nd}}$ $\boxed{\text{QUIT}}$ to return to the home screen.

- Calculate $f(5)$, which is now $y1(5)$, using the evalF command. The syntax for the command is evalF(dependent variable, independent variable, value for independent variable). Here are the keystrokes.
  Press $\boxed{\text{2nd}}$ $\boxed{\text{CALC}}$ (same key as $\boxed{\div}$) $\boxed{\text{F1}}$ for evalF.
  On the TI-86, press $\boxed{\text{2nd}}$ $\boxed{\text{CATLG-VARS}}$ $\boxed{\text{MORE}}$ $\boxed{\text{F4}}$ for EQU; on the TI-85, press $\boxed{\text{2nd}}$ $\boxed{\text{VARS}}$ (same key as $\boxed{3}$) $\boxed{\text{MORE}}$ $\boxed{\text{F3}}$ for EQU.
  If necessary, use the arrow keys to move the triangle so that it points toward $y1$. Then press $\boxed{\text{ENTER}}$.
  Press $\boxed{,}$ $\boxed{x\text{-VAR}}$ $\boxed{,}$ $\boxed{5}$ $\boxed{)}$ $\boxed{\text{ENTER}}$. You should get 98.

- Compute $f(8)$, which is now $y1(8)$, as follows.
  Press $\boxed{\text{2nd}}$ $\boxed{\text{ENTRY}}$ (same key as $\boxed{\text{ENTER}}$).
  Move the cursor over 5 and then press $\boxed{8}$ $\boxed{\text{ENTER}}$. You should get 233.

Here's another method for computing $y1(5)$ on the TI-86:
  Press $\boxed{\text{2nd}}$ $\boxed{\text{CATLG-VARS}}$ $\boxed{\text{MORE}}$ $\boxed{\text{F4}}$ for EQU. If necessary, use the arrow keys to move the triangle so that it points toward $y1$ and press $\boxed{\text{ENTER}}$. Press $\boxed{(}$ $\boxed{5}$ $\boxed{)}$ $\boxed{\text{ENTER}}$. You should get 98, the same answer that you got using evalF.

Do <u>not</u> erase the formula for $f(x)$ until you have completed the next example.

Example: Find the average rate of change of $f(x) = 3x^2 + 6x - 7$ from $x = 5$ to $x = 7$.
  From the previous example, you should already have the formula for $f(x)$ stored as $y1$. You perform these calculations from the home screen.

On the TI-86, enter the formula for the average rate of change and then press $\boxed{\text{ENTER}}$. Your screen should be similar to Figure 13.

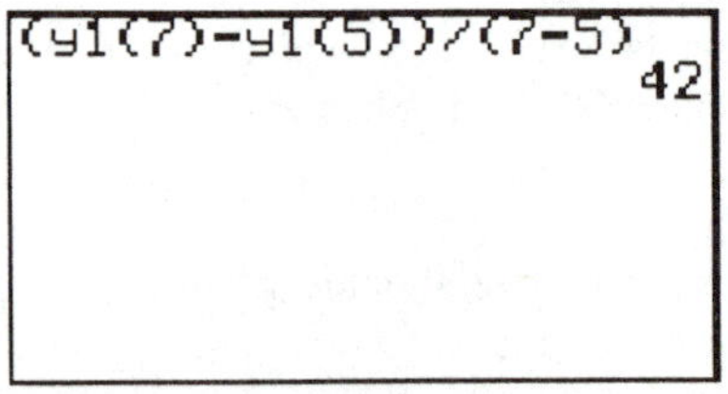

```
(y1(7)-y1(5))/(7-5)
                   42
```

Figure 13. Using the TI-86 to Calculate
an Average Rate of Change

On the TI-85, you will have to use evalF($y1,x,7$) in place of $y1(7)$ and evalF($y1,x,5$) in place of $y1(5)$. Your screen should be similar to Figure 14

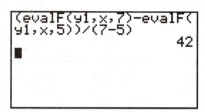

Figure 14. Using the TI-85 to Calculate
an Average Rate of Change

Adjusting the Window Settings for Square Scaling

The viewing screen on your calculator is a rectangle. Therefore, if you use the standard window, the tick marks on the y-axis will be closer together than those on the x-axis. For square scaling, you want the distance between 0 and 1 on the x-axis to be the same as the distance between 0 and 1 on the y-axis.

Example: The graph of $f(x) = \sqrt{25 - x^2}$ forms the top half of a circle. Graph $f(x) = \sqrt{25 - x^2}$, first in the standard window, and then switch to square scaling.
   You'll do this problem in two steps. In Step 1, you'll enter the function and graph it in the standard viewing window. In Step 2, you'll change to square scaling.

Step 1: Enter $\sqrt{25 - x^2}$ as $y1$ and then graph it in the standard viewing window.
   • Press $\boxed{\text{GRAPH}}$ $\boxed{\text{F1}}$. Erase any previously stored functions.

   • Enter $\sqrt{25 - x^2}$ as $y1$: press $\boxed{\text{2nd}}$ $\boxed{\sqrt{}}$ $\boxed{(}$ $\boxed{2}$ $\boxed{5}$ $\boxed{-}$ $\boxed{\text{F1}}$ $\boxed{x^2}$ $\boxed{)}$.

   • Press $\boxed{\text{EXIT}}$ and then $\boxed{\text{F3}}$ $\boxed{\text{F4}}$ for ZSTD. Observe the spacing between the tick marks on the x- and y-axes. Notice that the tick marks on the y-axis are closer together than the tick marks on the x-axis. This graph does not appear to be a semicircle.

   • Press $\boxed{\text{F2}}$ to view the window settings of the standard window.

Step 2: Change to square scaling.
   • Press $\boxed{\text{F3}}$ for ZOOM $\boxed{\text{MORE}}$ $\boxed{\text{F2}}$ to select ZSQR. When the graph appears on your screen, observe the equal distance between tick marks on the two axes. In this window, the graph is shaped like a semicircle.

   • Press $\boxed{\text{F2}}$ and note the changes in the window settings.

Continue directly to the next example.

Example: Add the graph of $g(x) = -\sqrt{25 - x^2}$ to the previous graph.

This function is the opposite of the previous function. So, you can define it in terms of $y1$. Here's how.

- Press $\boxed{\text{GRAPH}}$ $\boxed{\text{F1}}$. Your screen should be similar to Figure 15. Press $\boxed{\text{ENTER}}$ to move the cursor, in this case a solid rectangle, opposite $y2$.

Figure 15. Contents of $y(x)$= Editor

- Press $\boxed{\text{(-)}}$ and then $\boxed{\text{F2}}$ $\boxed{\text{1}}$ to create the function $-y1$.

- To see the graph of both halves of a circle, press $\boxed{\text{EXIT}}$ $\boxed{\text{F5}}$. Your graph should be similar to the one in Figure 16.

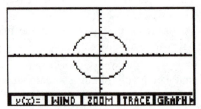

Figure 16. Graph of a Circle

- Press $\boxed{\text{2nd}}$ $\boxed{\text{QUIT}}$ to return to the home screen.

64

CHAPTER 2

Finding Values of a Recursive Function

Example: Suppose that $P(0) = 2$ and that $P(t + 1) = P(t) + 5$. Use your calculator to find $P(1)$, $P(2)$, $P(3)$, and $P(4)$.

There are several ways to solve this problem. Here's one that uses your calculator's answer feature.

- Enter the value for $P(0)$: press $\boxed{2}$ $\boxed{\text{ENTER}}$.

- Since $P(1) = P(0) + 5$, press $\boxed{+}$ $\boxed{5}$ $\boxed{\text{ENTER}}$ to add 5 to the value of $P(0)$.

- Press $\boxed{\text{ENTER}}$ three more times to find the values of $P(2)$, $P(3)$, and $P(4)$. (Check that your answers match those in Figure 17.)

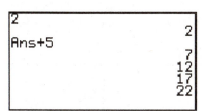

Figure 17. The Answer Feature

Fitting a Line to Data

If your data, when plotted, lie exactly on a line, you can use algebra to determine the equation of the line. However, real data seldom fall precisely on a line. Instead, the plotted data may exhibit a roughly linear pattern. The least squares line (also called the regression line) is a line that statisticians frequently use when describing a linear trend in data.

Example: Use the least squares line to describe the linear pattern in the data displayed in Table 3.

x	y
-3.0	-6.3
-2.0	-2.8
1.2	2.0
2.0	4.1
3.1	5.0
4.2	7.2

Table 3. Data With Linear Trend

You'll tackle this problem in three steps. In Step 1, you'll enter the data. In Step 2, you'll determine the equation of the least squares line. Then in Step 3, you'll plot the data and graph the line.

Step 1: Erase any stored functions and then enter the data.

- Press GRAPH F1 and erase any stored functions. Return to the home screen.

- Erase any stored data and then enter the data from Table 3. (See *Plotting Points*, Chapter 1, Steps 2 and 3, starting on page 57.) After entering your data, press 2nd QUIT to return to the home screen.

Step 2: Next, you'll find the equation for the least squares line. Your calculator will determine the values for the slope and intercept of $y = a + bx$. Here b represents the slope of the line, and a is the y-intercept. (Note: Mathematicians generally use b to represent the y-intercept. In this case, however, b represents the slope and not the y-intercept.)

On the TI-86:

- From the home screen, press 2nd STAT F1 for CALC.

- Press F3 for LinR, which is short for linear regression. Then press 2nd LIST (same key as −) F3 to access the names of your lists.

- Press F2 , F3 ENTER .

On the TI-85:

- From the home screen, press STAT F1 for CALC.

- Press ENTER twice to enter the STAT/CALC menu. Then press F2 for LINR, which is short for linear regression.

If you have entered your data correctly, your screen will be similar to the one in Figure 18.

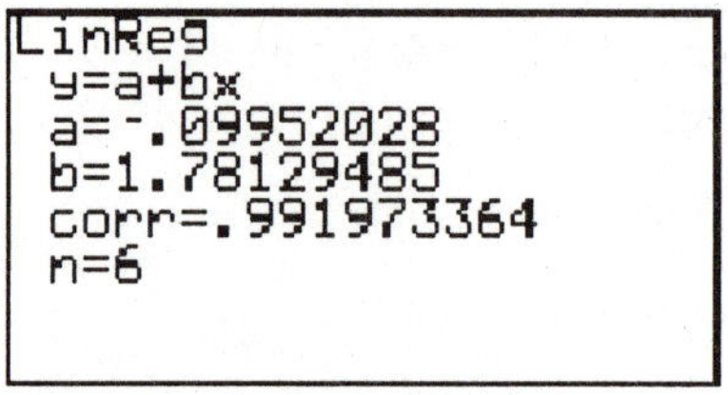

```
LinReg
y=a+bx
a=-.09952028
b=1.78129485
corr=.991973364
n=6
```

Figure 18. Output From LINR Command

The output from the LINR command tells you the values for a and b, the y-intercept and slope, respectively.

Step 3: Make a scatterplot of the data and overlay a graph of the least squares line.

- Press 2nd QUIT to return to the home screen. Then press GRAPH F1 . The cursor should be opposite $y1$.

• Insert the equation for the least squares line. Here's how:

On the TI-86:
      Press ⏐ 2nd ⏐⏐ STAT ⏐⏐ F5 ⏐ for VARS.
      Press ⏐ MORE ⏐⏐ MORE ⏐⏐ F2 ⏐ to select RegEq.

On the TI-85:
      Press ⏐ 2nd ⏐⏐ VARS ⏐ (same key as ⏐ 3 ⏐) ⏐ MORE ⏐⏐ MORE ⏐⏐ F3 ⏐ for STAT.
      Use the down arrow key to move the triangle opposite RegEq and press ⏐ ENTER ⏐.

You should see $y1 = $ RegEq at the top of your screen.

• Now graph the least squares line and the scatterplot of the data as follows:

On the TI-86:
      Press ⏐ 2nd ⏐⏐ QUIT ⏐ to return to the home screen.
      Press ⏐ 2nd ⏐⏐ STAT ⏐⏐ F3 ⏐ for PLOT.
      Press ⏐ F1 ⏐⏐ ENTER ⏐ to turn Plot 1 on.
      Press ⏐ GRAPH ⏐⏐ F3 ⏐⏐ MORE ⏐⏐ F5 ⏐ for ZDATA.
      Press ⏐ CLEAR ⏐ to remove the menu bar from the bottom of your screen. Your graph
          should be similar to the one in Figure 19.

On the TI-85:
      Press ⏐ EXIT ⏐ to return to the GRAPH menu.
      Press ⏐ F2 ⏐ and adjust the window settings so that all data points, when plotted, will
          appear on your screen.
      Press ⏐ STAT ⏐⏐ F3 ⏐ for DRAW. A graph of the least squares line will appear on your
          screen.
      Press ⏐ F2 ⏐ to add the scatterplot to the display. Then press ⏐ CLEAR ⏐. Your graph
          should be similar to the one in Figure 19.

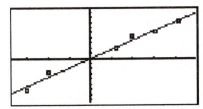

Figure 19. Scatterplot and Regression Line

• Turn off the scatterplot.

On the TI-86, press ⏐ 2nd ⏐⏐ STAT ⏐⏐ F3 ⏐⏐ F5 ⏐⏐ ENTER ⏐ to turn off Plot 1.

On the TI-85, press ⏐ STAT ⏐⏐ F3 ⏐⏐ F5 ⏐ for CLDRW, which is short for clear draw.

Note: *You can adapt the previous instructions to fit other functions to data. For example, if you want to fit an exponential function (instead of a linear function), select ExpR (instead of LinR) from the STAT/CALC menu. P2Reg will fit a quadratic function, P3Reg will fit a cubic function.*

Graphing an Exponential Function

Example: Graph $p(x) = 200(1.09)^x$ over the interval from $x = 0$ to $x = 10$. Use zoom fit to choose the window settings for yMin and yMax. Then change the window settings and graph the function over the interval from $x = 0$ to $x = 50$

- Press GRAPH F1 and clear any stored functions.

- Enter the formula for $p(x)$ as $y1$. (Use the ^ key for the exponent.)

- Change the window settings as follows: press EXIT to return to the GRAPH menu. Then press F2 to access the window settings. Set xMin = 0, xMax = 10, and yScl = 0 (to turn off the tick marks on the y-axis).

- Next, let your calculator set the values for yMin and yMax. Here's how.
  Press F3 for ZOOM.
  Press MORE F1 for ZFIT, which is short for zoom fit. Wait a few seconds to view the graph.
  Press CLEAR to remove the menu bar from the bottom of the screen. Your screen should be similar to the one in Figure 20.

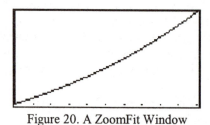

Figure 20. A ZoomFit Window

- Check the window settings: press GRAPH F2 .

- Change the window settings for $x$max to 50. Then press F3 MORE F1 for ZFIT. In this window, the graph should appear to curve upward more steeply.

Finding the Coordinates of Points of Intersection

Example: Find the points where the graphs of $f(x) = -4x + 15$ and $g(x) = -2x^2 + 2x + 12$ intersect. You'll tackle this problem in three steps. In Step 1, you'll graph the functions. You'll find the coordinates of one of the points of intersection in Step 2, and the other in Step 3.

Step 1: Graph the functions in a viewing window that gives a clear view of the points of intersection.
- Press $\boxed{\text{GRAPH}}$ $\boxed{\text{F1}}$. Erase any stored functions and then enter the $f(x)$ and $g(x)$ as $y1$ and $y2$ respectively.

- Adjust the window settings so that you can see both points of intersection. (*Hint:* you might start with the standard viewing window. From the GRAPH menu, press $\boxed{\text{F3}}$ $\boxed{\text{F4}}$. Then press $\boxed{\text{F2}}$ and adjust the window settings after looking at the graph.)

Step 2: Approximate the coordinates of one of the points of intersection.
- From the GRAPH menu, press $\boxed{\text{MORE}}$ $\boxed{\text{F1}}$ to access the GRAPH/MATH menu.

- Press $\boxed{\text{MORE}}$ again. Next press the F-key that corresponds to ISECT($\boxed{\text{F3}}$ on the TI-86 and $\boxed{\text{F5}}$ on the TI-85). Your screen should be similar to Figure 21.

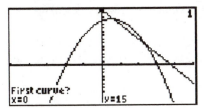

Figure 21. Intersecting Curves

- Your calculator wants to know which curve you want to call the first curve. (The TI-85 simply shows a 1 in the upper right-hand corner.) Try pressing the up and down arrow keys. The cursor will jump back and forth from the line to the parabola. Position the cursor on the line. (We'll designate the line as the first curve since its equation is entered as $y1$.) Then press $\boxed{\text{ENTER}}$. The cursor should then jump to the parabola.

  On the TI-86, press $\boxed{\text{ENTER}}$ to designate the parabola as the second curve. The TI-85 assumes that the parabola is the second curve, so skip directly to the next bullet without pressing $\boxed{\text{ENTER}}$

- Your calculator wants a guess. Use the right or left arrow keys to position the cursor (a box with a blinking $\times$) on top of the intersection point with the smaller x-coordinate. Then press $\boxed{\text{ENTER}}$. Read the coordinates of this intersection point from the bottom of your screen.

Step 3: Find the coordinates of the other intersection point.
Repeat the process outlined in Step 2 to find the coordinates of the other intersection point.

If you have done everything correctly, you will find that the two graphs intersect at approximately (0.634, 12.464) and (2.366, 5.536).

Graphing Piecewise-Defined Functions

Example: Graph the piecewise-defined function $f(x) = \begin{cases} x - 4, & \text{if } x > 4 \\ -x + 4, & \text{if } x \leq 4 \end{cases}$.

The graph of $f(x)$ consists of two half-lines pieced together. You'll want the graph of $y = x - 4$ when x-values are greater than 4 and $y = -x + 4$ when x-values are less than or equal to 4.

You'll complete this example in three steps. In Step 1, you'll enter two linear functions and in Step 2, you'll piece them together. In Step 3, you'll graph $f(x)$.

Step 1: Enter the functions that you want to piece together.
- Press ⎡GRAPH⎤ ⎡F1⎤ and clear any previously stored functions. Then enter $x - 4$ as $y1$ and $-x + 4$ as $y2$.

- Press ⎡EXIT⎤ ⎡F3⎤ ⎡F4⎤. Your graph should look like a cross: $\times$.

Step 2: Piece together the functions entered in Step 1 to create $f(x)$.
- Return to the GRAPH/$y(x) =$ menu. Press ⎡ENTER⎤ twice to position the cursor opposite $y3$.

- To insert $y1$: press ⎡F2⎤ ⎡1⎤.

- Press ⎡ $\times$ ⎤. Enter the condition that governs when to use $y1$: press ⎡(⎤ ⎡F1⎤ ⎡2nd⎤ ⎡TEST⎤ (same key as ⎡2⎤) ⎡F3⎤ ⎡4⎤ ⎡)⎤. Press ⎡EXIT⎤ to return to the GRAPH/$y(x) =$ menu.

- Press ⎡+⎤.

- To insert $y2$: press ⎡F2⎤ ⎡2⎤.

- Press ⎡ $\times$ ⎤. Enter the condition that governs when to use $y2$: press ⎡(⎤ ⎡F1⎤ ⎡2nd⎤ ⎡TEST⎤ ⎡F4⎤ ⎡4⎤ ⎡)⎤ ⎡ENTER⎤. When you have completed this step, your screen should be similar to Figure 22.

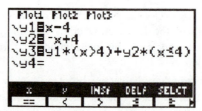

Figure 22. Entering the Formula for $f(x)$

Here's how your calculator interprets the information you've just entered as a piecewise-defined function. The calculator assigns the expression $(x > 4)$ the value 1 when the inequality is true (in other words, when the input variable , x, is greater than 4). In this case, the inequality $x \leq 4$ is false, so the calculator sets the expression $x \leq 4$ equal to 0. Thus, for $x > 4$, the function $y3$ is equivalent to:

$$y3 = (x - 4)(1) + (-x + 4)(0) = x - 4.$$

And when $x \leq 4$, the function $y3$ is equivalent to:

$$y3 = (x - 4)(0) + (-x + 4)(1) = -x + 4.$$

Step 3: Graph $f(x)$.
- Press $\boxed{\text{EXIT}}$ to return to the GRAPH/$y(x) =$ menu.

- Unselect $y1$ and $y2$:
   Move the cursor opposite $y1$. Press $\boxed{\text{F5}}$ to remove the highlighting from $y1$'s equals sign.
   Then position the cursor opposite $y2$ and press $\boxed{\text{F5}}$.

- Now, press $\boxed{\text{EXIT}}$ $\boxed{\text{F5}}$. The graph of $f(x)$ should look V-shaped.

Example: Graph $g(x) = \begin{cases} -2x, & \text{if } x < 1 \\ -2 + 3(x - 1), & \text{if } 1 \leq x < 3 \\ 4, & \text{if } x \geq 3 \end{cases}$.

In the previous example, you entered separate functions and then pieced them together to make a new single function. This time, you'll enter the three-piece function as $y1$. There is one problem: your calculator is not sophisticated enough to understand the condition $1 \leq x < 3$. Instead you'll have to enter the equivalent condition $1 \leq x$ and $x < 3$.

- Press $\boxed{\text{GRAPH}}$ $\boxed{\text{F1}}$ and erase any stored functions.

- In the GRAPH/$y(x) =$ editor, enter $g(x)$ as $y1$:
   Press $\boxed{(-)}$ $\boxed{2}$ $\boxed{x\text{-VAR}}$.
   Press $\boxed{\times}$ $\boxed{(}$ $\boxed{x\text{-VAR}}$ $\boxed{\text{2nd}}$ $\boxed{\text{TEST}}$ $\boxed{\text{F2}}$ for <, and then press $\boxed{1}$ $\boxed{)}$.
   Press $\boxed{+}$.
   Press $\boxed{(}$ $\boxed{(-)}$ $\boxed{2}$ $\boxed{+}$ $\boxed{3}$ $\boxed{(}$ $\boxed{x\text{-VAR}}$ $\boxed{-}$ $\boxed{1}$ $\boxed{)}$ $\boxed{)}$.
   Press $\boxed{\times}$ $\boxed{(}$ $\boxed{1}$. You should still be in the test menu. Press $\boxed{\text{F4}}$ for $\leq$ and then $\boxed{x\text{-VAR}}$.
   To insert *and*: press $\boxed{\text{2nd}}$ $\boxed{\text{BASE}}$ (same key as $\boxed{1}$) $\boxed{\text{F4}}$ to enter the BASE/BOOL menu.
      Press $\boxed{\text{F1}}$ for *and*.
   Press $\boxed{x\text{-VARS}}$. Press $\boxed{\text{2nd}}$ $\boxed{\text{TEST}}$ to return to the TEST menu and press $\boxed{\text{F2}}$ for <. Then
      press $\boxed{3}$ $\boxed{)}$.
   Press $\boxed{+}$.
   Press $\boxed{4}$ $\boxed{\times}$ $\boxed{(}$ $\boxed{x\text{-VAR}}$. You should still be in the TEST menu. Press $\boxed{\text{F5}}$ for $\geq$ and
      then press $\boxed{3}$ $\boxed{)}$. Finally, press $\boxed{\text{ENTER}}$.

• Press ⎡EXIT⎤ twice to return to the GRAPH menu. Then press ⎡F3⎤ ⎡F4⎤ to graph $g(x)$. If you entered the function correctly, your graph should match the one shown in Figure 23.

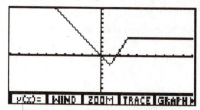

Figure 23. The Graph of $g(x)$

Example: Graph the greatest integer function, $[[x]]$ in the standard viewing window. (See Chapter 2, Exercise 46 in *Precalculus: Concepts in Context, 2e* for a definition of this function.)

The greatest integer function is one of your calculator's built-in functions. Its calculator name is $int(x)$. You can access it from the catalog.

• Press ⎡GRAPH⎤ ⎡F1⎤ and clear any stored functions. Position the cursor opposite $y1$.

• To enter the function, press ⎡2nd⎤ ⎡CATALOG⎤ (same key as ⎡CUSTOM⎤). On a TI-86, you will need to press ⎡F1⎤ to access the catalog. To move to the i's Press ⎡I⎤ (same key as ⎡)⎤). Press the down arrow key until the triangle points to **int**. Then press ⎡ENTER⎤ ⎡(⎤ ⎡F1⎤ ⎡)⎤.

• Press ⎡EXIT⎤ ⎡F3⎤ ⎡F4⎤ to view the graph. Your graph should resemble Figure 24.

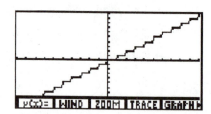

Figure 24. Graph of the Greatest Integer Function

The vertical line segments that appear on your screen cannot be part of the graph of a function. (Why?) For example, the nearly vertical line segment at $x = 3$ is due to the fact that your graphing calculator plotted a point on the left side of $x = 3$, and a point on the right of $x = 3$, and then connected the points with a straight line segment. One solution to this problem is to change the format from DRAWLINE to DRAWDOT. That's what you'll do next.

• You should still be in the GRAPH menu. Press ⎡MORE⎤ ⎡F3⎤ for FORMT. Change the setting from DRAWLINE to DRAWDOT. Then press ⎡F5⎤ to view the graph. The nearly vertical line segments should now be gone.

• Press ⎡MORE⎤ ⎡F3⎤ and return the setting to DRAWLINE. Return to the home screen.

CHAPTER 3

In Chapter 3 you will be investigating the effect that certain algebraic modifications, such as adding a constant to the input variable, have on the graph of a function. You'll want to experiment using several different functions. We've provided some functions and algebraic modifications that you might want to consider.

Using Parentheses

Warning! *When you want to apply a function to an expression, you must enclose the entire expression in parentheses.*

Example: Graph $y = \sqrt{x + 2}$ in the window $[-5, 5] \times [-5, 5]$.

- Press $\boxed{\text{GRAPH}}$ $\boxed{\text{F1}}$ and erase any previously stored functions. Then enter $y = \sqrt{x + 2}$ by pressing $\boxed{\text{2nd}}$ $\boxed{\sqrt{}}$ $\boxed{(}$ $\boxed{\text{F1}}$ $\boxed{+}$ $\boxed{2}$ $\boxed{)}$.

- Now, press $\boxed{\text{EXIT}}$ $\boxed{\text{F2}}$ and set xMin $= -5$, xMax $= 5$, yMin $= -5$ and yMax $= 5$. Press $\boxed{\text{F5}}$ to graph this function in the $[-5, 5] \times [-5, 5]$ window. Your graph should resemble the one in Figure 25.

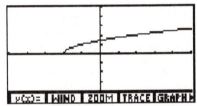

Figure 25. A Member of the Square Root Family

Graphing Functions Involving Absolute Value

Example: Graph $y = |x|$. Then add the graph of $g(x) = \frac{|x|}{x}$. Use settings for a $[-5, 5] \times [-5, 5]$ window. You can access the absolute value function from the catalog.

- Press $\boxed{\text{GRAPH}}$ $\boxed{\text{F1}}$ and clear any stored functions.

- Enter $y = |x|$ as $y1$: press $\boxed{\text{2nd}}$ $\boxed{\text{CATALOG}}$ (same key as $\boxed{\text{CUSTOM}}$). On the TI-86, you will need to press $\boxed{\text{F1}}$ to access the catalog. Press $\boxed{\text{A}}$ (same key as $\boxed{\text{LOG}}$) to move to the A's. The triangle should be pointing to **abs**. Press $\boxed{\text{ENTER}}$ followed by $\boxed{(}$ $\boxed{\text{F1}}$ $\boxed{)}$.

- Press $\boxed{\text{EXIT}}$ to return to the GRAPH menu. If you worked through the previous example, you already have the correct window settings. (Otherwise press $\boxed{\text{F2}}$ and adjust the window settings.) Press $\boxed{\text{F5}}$ to view the V-shaped graph of the absolute value function.

73

- Next, enter the formula for $g(x)$ as $y2$:

    From the GRAPH menu, press $\boxed{\text{F1}}$.

    Press $\boxed{\text{ENTER}}$ to move the cursor opposite $y2$.

    Press $\boxed{\text{2nd}}$ $\boxed{\text{CATALOG}}$ (same key as $\boxed{\text{CUSTOM}}$). On the TI-86, press $\boxed{\text{F1}}$ to access the catalog.

    With the triangle pointing to **abs**, press $\boxed{\text{ENTER}}$. Then press $\boxed{\text{(}}$ $\boxed{\text{F1}}$ $\boxed{\text{)}}$ $\boxed{\div}$ $\boxed{\text{F1}}$.

- Press $\boxed{\text{EXIT}}$ to return to the GRAPH menu and then $\boxed{\text{F5}}$ to view the two graphs.

Note: *The function $g(x) = \frac{|x|}{x}$ tells you the sign of the input. The value of $g(x)$ is -1 for negative inputs and $+1$ for positive inputs. It is undefined at zero.*

Do <u>not</u> clear $y1$ and $y2$ from your calculator until you have completed the next example.

Example: Use your calculator to evaluate $f(x) = |x|$ and $g(x) = \frac{|x|}{x}$ at $x = 0$.

From the previous example, you should already have $f(x)$ and $g(x)$ stored as $y1$ and $y2$, respectively. Begin these calculations from the home screen.

- First, evaluate $f(0)$:

    Press $\boxed{\text{2nd}}$ $\boxed{\text{CALC}}$ $\boxed{\text{F1}}$ for evalF.

    Press $\boxed{\text{2nd}}$ $\boxed{\text{VARS}}$ $\boxed{\text{MORE}}$ and then press the F-key corresponding to EQU.

    Move the triangle so that it points to $y1$ and press $\boxed{\text{ENTER}}$.

    Press $\boxed{,}$ $\boxed{x\text{-VAR}}$ $\boxed{,}$ $\boxed{0}$ $\boxed{)}$ $\boxed{\text{ENTER}}$. You should get 0.

Next, evaluate $g(0)$:

    Press $\boxed{\text{2nd}}$ $\boxed{\text{ENTRY}}$.

    Move the cursor over y.

    Press $\boxed{\text{2nd}}$ $\boxed{\text{VARS}}$ $\boxed{\text{MORE}}$ and then press the F-key corresponding to EQU.

    Move the triangle so that it points to $y2$ and press $\boxed{\text{ENTER}}$.

    Press $\boxed{\text{ENTER}}$. Do you see why $g(x)$ is not defined at $x = 0$?

Note: *If you modify g(x) by setting g(0) = 0, then you get the sign function. You can access the sign function from your calculator's catalog. This function is defined by the following piecewise formula:*

$$\text{sign}(x) = \begin{cases} \frac{|x|}{x}, & \text{if } x \neq 0 \\ 0, & \text{if } x = 0 \end{cases}.$$

The sign function is not to be confused with the sine function, the next topic.

Graphing the Sine Function

Example: Graph $y = \sin(x)$ in the trigonometric viewing window.
  You will learn more about this function in Chapter 6.

- First check that your calculator is in Radian Mode. Press $\boxed{\text{2nd}}$ $\boxed{\text{MODE}}$. If Radian is not highlighted, move the cursor over Radian and press $\boxed{\text{ENTER}}$. Press $\boxed{\text{2nd}}$ $\boxed{\text{QUIT}}$ to return to the home screen.

- Press $\boxed{\text{GRAPH}}$ $\boxed{\text{F1}}$ and erase any stored functions. The cursor should be opposite $y1 = $. Press $\boxed{\text{SIN}}$ $\boxed{(}$ $\boxed{\text{F1}}$ $\boxed{)}$.

- Press $\boxed{\text{EXIT}}$ $\boxed{\text{F3}}$ $\boxed{\text{MORE}}$ $\boxed{\text{F3}}$ for ZTRIG to view the graph of $y = \sin(x)$ in the trig viewing window.

Graphing a Family of Functions

Using your calculator's list capabilities, you can substitute each value in a given list for a constant in an algebraic formula. This feature allows you to graph an entire family of functions quickly. On the TI-85/86 you specify a list by enclosing the members of the list in brackets: { }.

Example: Graph the family of quadratic functions $y = (x + 1)^2$, $y = (x + 2)^2$, and $y = (x + 3)^2$ in the window $[-5, 5] \times [-1, 10]$.
- From the home screen, press $\boxed{\text{GRAPH}}$ $\boxed{\text{F2}}$ and adjust the settings for a $[-5, 5] \times [-1, 10]$ window.

- You should still be in the GRAPH menu. Press $\boxed{\text{F1}}$ and clear any stored functions. If necessary, move the cursor opposite $y1$.

- Enter the three functions by specifying the constants 1, 2, and 3 in a list as follows:
  Press $\boxed{(}$ $\boxed{\text{F1}}$ $\boxed{+}$
  Press $\boxed{\text{2nd}}$ $\boxed{\text{LIST}}$ (same key as $\boxed{-}$) $\boxed{\text{F1}}$ for { and then press $\boxed{1}$ $\boxed{,}$ $\boxed{2}$ $\boxed{,}$ $\boxed{3}$.
  You should still be in the LIST menu. Press $\boxed{\text{F2}}$ for }.
  Press $\boxed{)}$ $\boxed{x^2}$.

- Press $\boxed{\text{EXIT}}$ twice to return to the GRAPH menu. Then press $\boxed{\text{F5}}$ and watch as the three functions are graphed one after the other.

Composing Functions

Example: Suppose $f(x) = 2x - 7$ and $g(x) = 5x^2$. Find the value of $f(g(5))$.
- Press $\boxed{\text{GRAPH}}$ $\boxed{\text{F1}}$ and erase any previously stored functions.

- Enter the formula for $f(x)$ as $y1$ and the formula for $g(x)$ as $y2$. Then press $\boxed{\text{2nd}}$ $\boxed{\text{QUIT}}$ or $\boxed{\text{EXIT}}$ twice to return to the home screen.

- Evaluate the value of the composition as follows:

  On the TI-86:
  > Press $\boxed{\text{2nd}}$ $\boxed{\text{CATLG-VARS}}$ $\boxed{\text{MORE}}$ $\boxed{\text{F4}}$ for EQU.
  > If necessary, move the triangle so that it points to $y1$ and press $\boxed{\text{ENTER}}$.
  > Press $\boxed{\,(\,}$.
  > Press $\boxed{\text{2nd}}$ $\boxed{\text{CATLG-VARS}}$ $\boxed{\text{MORE}}$ $\boxed{\text{F4}}$.
  > Move the triangle so that it points to $y2$ and press $\boxed{\text{ENTER}}$.
  > Press $\boxed{\,(\,}$ $\boxed{5}$ $\boxed{\,)\,}$ $\boxed{\,)\,}$ $\boxed{\text{ENTER}}$.

  On the TI-85, you will need to nest evalF commands. Here's how:
  > Press $\boxed{\text{2nd}}$ $\boxed{\text{CALC}}$ $\boxed{\text{F1}}$ for evalF.
  > Press $\boxed{\text{2nd}}$ $\boxed{\text{VARS}}$ $\boxed{\text{MORE}}$ $\boxed{\text{F3}}$ for EQU.
  > If necessary, move the triangle so that it points to $y1$ and press $\boxed{\text{ENTER}}$.
  > Press $\boxed{\,,\,}$ $\boxed{x\text{-VAR}}$ $\boxed{\,,\,}$.
  > Press $\boxed{\text{2nd}}$ $\boxed{\text{CALC}}$ $\boxed{\text{F1}}$.
  > Press $\boxed{\text{2nd}}$ $\boxed{\text{VARS}}$ $\boxed{\text{MORE}}$ $\boxed{\text{F3}}$.
  > Move the triangle so that it points to $y2$ and press $\boxed{\text{ENTER}}$.
  > Press $\boxed{\,,\,}$ $\boxed{x\text{-VAR}}$ $\boxed{\,,\,}$ $\boxed{5}$ $\boxed{\,)\,}$ $\boxed{\,)\,}$ $\boxed{\text{ENTER}}$.

The value of the composition at $x = 5$ should be 243.

Do <u>not</u> erase $f(x)$ and $g(x)$ until after you have completed the next example.

Example: Given $f(x) = 2x - 7$ and $g(x) = 5x^2$. Graph $f(g(x))$ in a window that shows the key features of the graph.

From the previous example, you should have $f(x)$ stored as $y1$ and $g(x)$ stored as $y2$. First, you'll form the composition and store it as $y3$. Then you'll graph it. (After completing this example, you may decide that it's easier to find the formula for the composition by hand and then graph that formula.)

- Press $\boxed{\text{GRAPH}}$ $\boxed{\text{F1}}$. Press $\boxed{\text{ENTER}}$ twice to move the cursor opposite $y3$.

- Enter the composition $f(g(x))$, which is now $y1(y2(x))$ as $y3$. Here's how:

  On the TI-86:
  > Press $\boxed{\text{2nd}}$ $\boxed{\text{CATLG-VARS}}$ $\boxed{\text{MORE}}$ $\boxed{\text{F4}}$ for EQU.
  > Move the triangle so that it points to $y1$ and press $\boxed{\text{ENTER}}$ and then $\boxed{\,(\,}$.
  > Press $\boxed{\text{2nd}}$ $\boxed{\text{CATLG-VARS}}$ $\boxed{\text{MORE}}$ $\boxed{\text{F4}}$.
  > Move the triangle so that it points to $y2$ and press $\boxed{\text{ENTER}}$.
  > Press $\boxed{\,(\,}$ $\boxed{x\text{-VAR}}$ $\boxed{\,)\,}$ $\boxed{\,)\,}$.

On the TI-85 the process is more complicated. You will need to nest evalF commands.

You might think of it this way. If $f(x)$ is stored as $y1$, then you can evaluate $f(5)$ by evalF($y1,x,5$) and $f(10)$ by evalF($y1,x,10$). For an arbitrary value of x, you can evaluate $f(x)$ by evalF($y1,x,x$). Hence, you can evaluate the composition, $f(g(x))$, or $y1(y2(x))$ by evalF($y1,x$, evalF($y2,x,x$)). Here are the keystrokes.

Press $\boxed{\text{2nd}}$ $\boxed{\text{CALC}}$ $\boxed{\text{F1}}$ for evalF.

Press $\boxed{\text{EXIT}}$ $\boxed{\text{F2}}$ $\boxed{1}$ to enter $y1$.

Press $\boxed{,}$ $\boxed{\text{x-VAR}}$ $\boxed{,}$.

Press $\boxed{\text{2nd}}$ $\boxed{\text{CALC}}$ $\boxed{\text{F1}}$ for evalF.

Press $\boxed{\text{EXIT}}$ $\boxed{\text{F2}}$ $\boxed{2}$ to enter $y2$.

Press $\boxed{,}$ $\boxed{\text{x-VAR}}$ $\boxed{,}$.

Press $\boxed{\text{x-VARS}}$ $\boxed{)}$ $\boxed{)}$

- Unselect the functions stored as $y1$ and $y2$:

  In the GRAPH/$y(x) =$ menu, position the cursor on the line containing $y1$. Then press $\boxed{\text{F5}}$. Repeat the process for $y2$.

- Press $\boxed{\text{EXIT}}$ to return to the GRAPH menu. Then press $\boxed{\text{F2}}$ and adjust the settings for a $[-5,5] \times [-5,5]$ window. Press $\boxed{\text{F5}}$ to view the graph. Your graph should look like a parabola.

CHAPTER 4

Graphing Exponential Functions Involving e

There are two bases for exponential functions that are so common that they have their own function keys on the TI-85/86: e^x and 10^x.

Example: Graph $f(x) = e^x$ in the window $[-3, 3] \times [-1, 12]$.
- Press GRAPH F1 and clear any stored functions. Then enter $f(x)$ as $y1$: press 2nd e^x (same key as LN) (F1) .

- Press EXIT to return to the GRAPH menu. Press F2 and adjust the settings for a $[-3, 3] \times [-1, 12]$ window.

- Press F5 to view the graph.

Continue directly to the next example.

Example: Graph $g(x) = e^{-\frac{x}{2}}$ by editing the function from the previous example.
- From the GRAPH menu, press F1 to access the GRAPH/$y(x) =$ menu. The cursor should be on the line containing the formula $y1=e^{\wedge}(x)$.

- Use the right arrow key to move the cursor over the x. Press 2nd INS (-) . Then use the right arrow key to move the cursor over the right parenthesis and press 2nd INS ÷ 2 .

- Press EXIT F5 to view the function in the same window as the previous example.

Do <u>not</u> erase $g(x)$ until you have worked through the next topic, *Turning Off the Axes*.

Turning Off the Axes

Sometimes it is helpful to view a graph without the presence of the x and y axes. In the last example, you may have noticed that the graph of $g(x)$ disappeared into the x-axis. Did the graph, in reality, disappear? In the next example, you'll turn off the axes and find out.

Example: View the graph of $g(x) = e^{-\frac{x}{2}}$ in the window $[-3, 3] \times [-1, 12]$ with the axes turned off. You should already have $g(x)$ stored as $y1$.

- From the GRAPH menu, press MORE F3 for FORMT.

- Change the setting from AxesOn to AxesOff. Then press F5 .

- Turn the axes back on in preparation for your next graph.

78

Graphing Logarithmic Functions

The natural logarithmic function (base e), $\ln(x)$, and the common logarithmic function (base 10), $\log(x)$, have their own function keys, $\boxed{\text{LN}}$ and $\boxed{\text{LOG}}$, respectively. Logarithmic functions of other bases can be graphed by dividing $\ln(x)$ or $\log(x)$ by the appropriate scaling factor.

Example: Graph $h(x) = \ln(x)$ and $g(x) = \log(x)$ in the window $[-1, 12] \times [-2, 3]$.
- From the home screen, press $\boxed{\text{GRAPH}}$ $\boxed{\text{F1}}$ and clear any previously stored functions.

- Enter $h(x)$ as $y1$: press $\boxed{\text{LN}}$ $\boxed{(}$ $\boxed{\text{F1}}$ $\boxed{)}$ $\boxed{\text{ENTER}}$.

- Enter $g(x)$ as $y2$: press $\boxed{\text{LOG}}$ $\boxed{(}$ $\boxed{\text{F1}}$ $\boxed{)}$.

- Press $\boxed{\text{EXIT}}$ to return to the GRAPH menu. Then press $\boxed{\text{F2}}$ and adjust the settings for a $[-1, 12] \times [-2, 3]$ window. Press $\boxed{\text{F5}}$ to view the two graphs.

Approximating Instantaneous Rates of Change

You can approximate the instantaneous rate of change of a function $f(x)$ at $x = a$ by evaluating the average rate of change of the function over a small interval containing a. For example, you might choose a symmetric interval about a of the form $(a - \delta, a + \delta)$ for some small value of δ. Using this interval, your approximation of the instantaneous rate of change would be

$$\frac{f(a + \delta) - f(a - \delta)}{(a + \delta) - (a - \delta)} = \frac{f(a + \delta) - f(a - \delta)}{2\delta}.$$

This is called the central difference quotient, which can be computed using the command **nDer**. The syntax of the command is shown below:

$$\text{nDer(function, variable, value)}$$

Example: Use the central difference quotient to approximate the instantaneous rate of change of the function $f(x) = e^{2x}$ at $x = 0, 1,$ and 2. On the TI-85, use $\delta = 0.01$.
    The TI-85 will let you choose the value for δ. The TI-86 will give an approximation to the instantaneous rate of change without your having to specify a value for δ.

- Press $\boxed{\text{GRAPH}}$ $\boxed{\text{F1}}$ and clear any previously stored functions.

- Enter e^{2x} for $y1$. (Be sure to enclose the exponent in parentheses.) Then press $\boxed{\text{2nd}}$ $\boxed{\text{QUIT}}$ to return to the home screen.

- Set the value for δ:
    If you are using a TI-86, skip directly to the next bullet. On the TI-85, set the value for δ as follows: press $\boxed{\text{2nd}}$ $\boxed{\text{TOLER}}$. Set $\delta = .01$. Press $\boxed{\text{2nd}}$ $\boxed{\text{QUIT}}$ to return to the home screen.

- Press $\boxed{\text{2nd}}$ $\boxed{\text{CALC}}$ (same key as $\boxed{\div}$) $\boxed{\text{F2}}$ for nDer.

- To complete the command:
  Press $\boxed{\text{2nd}}$ $\boxed{\text{VARS}}$ (on the TI-86, same key as $\boxed{\text{CUSTOM}}$; on the TI-85, same key as $\boxed{3}$)
  Press $\boxed{\text{MORE}}$ and then the F-key that corresponds to EQU.
  Move the triangle so that it points to $y1$ and press $\boxed{\text{ENTER}}$.
  Press $\boxed{,}$ $\boxed{x\text{-VAR}}$ $\boxed{,}$ $\boxed{0}$ $\boxed{)}$ $\boxed{\text{ENTER}}$. Your answer should be approximately 2.0.

- Edit the previous command to approximate the rate of change at $x = 1$. Press $\boxed{\text{2nd}}$ $\boxed{\text{ENTRY}}$.
  Change the value 0 to 1 and press $\boxed{\text{ENTER}}$. Repeat the process again, this time changing the value 1 to 2. You should get approximately 14.8 and 109.2, respectively.

CHAPTER 5

Approximating the Zeros of a Polynomial Function

Example: Find the zeros of $f(x) = 6x^3 - 49x^2 + 50x + 168$. In other words, find the x-values that make the function zero.

    To find the zeros of a polynomial function, locate the x-intercepts of its graph.

- Press ⌑GRAPH⌑ ⌑F1⌑ and clear any stored functions. Enter the formula for $f(x)$ as $y1$.

- Press ⌑EXIT⌑ ⌑F3⌑ ⌑F4⌑ to view the graph in the standard viewing window. Notice that this is not a good window for seeing the key features of this function. You'll need to adjust the window settings so that you can clearly see all turning points and x- and y-intercepts. In the GRAPH menu, press ⌑F2⌑ and change the window settings to match those in Figure 26. Then press ⌑F5⌑. Your graph should look like the one in Figure 27.

Figure 26. Window Settings   Figure 27. Graph of $f(x)$

Notice that the graph crosses the x-axis in three places. First, find the smallest x-intercept (the one that is negative).

- You should still be in the GRAPH menu. Press ⌑MORE⌑ ⌑F1⌑ to access the GRAPH/MATH menu.

- Approximate the negative x-intercept. To do so, you'll need to specify a small x-interval around the intercept and then an initial guess. Here's how:

    On the TI-86:
        Press ⌑F1⌑ for ROOT.
        Use the left arrow key to position the cursor (a box with a blinking $\times$) to the left of this x-intercept, and press ⌑ENTER⌑ to mark with a black triangle the lower bound of this interval.
        Use the right arrow key to position the cursor slightly to the right of this x-intercept (*but still to the left of the positive intercepts*) and press ⌑ENTER⌑, this time marking the upper bound for the interval.
        Use the right or left arrow keys to move the cursor close to this x-intercept. (Your screen should be similar to Figure 28.) Press ⌑ENTER⌑ and read the x-value from the bottom of your screen.

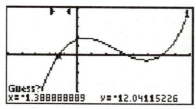

Figure 28. Finding the Zeros of $f(x)$

On the TI-85:

In the GRAPH/MATH menu, press $\boxed{\text{F1}}$ for LOWER.

Use the left arrow key to position the cursor (a + with a blinking center pixel) to the left of this x-intercept. (If you can't see the cursor, press the up arrow key to move it up slightly.) Then press $\boxed{\text{ENTER}}$ to mark with a black triangle the lower bound of this interval.

Press $\boxed{\text{F2}}$ for UPPER. Move the cursor slightly to the right of this x-intercept (*but still to the left of the positive intercepts*) and press $\boxed{\text{ENTER}}$, this time marking the upper bound for the interval.

Press $\boxed{\text{F3}}$ for ROOT. Move the cursor (a box with a blinking $\times$) close to the negative x-intercept. (Your screen should be similar to Figure 28.) Press $\boxed{\text{ENTER}}$ and read the x-value from the bottom of your screen.

• Repeat the process to find the values of the two positive zeros.

The zeros of $f(x)$ are $-\frac{4}{3}$ (≈ -1.33), 6, and 3.5.

Finding Local Maxima or Minima

Example: Estimate the local maximum and local minimum of $f(x) = x^3 - 4x^2 + 2x - 4$.

You'll complete this example in three steps. In Step 1, you'll graph $f(x)$. In Step 2, you'll find the local minimum. In Step 3, you'll find the local maximum.

Step 1: Graph $f(x)$ using a window that gives you a clear view of the two turning points (one peak and one valley) of the graph.

• Press $\boxed{\text{GRAPH}}$ $\boxed{\text{F1}}$, clear any previously stored functions, and then enter $f(x)$ as $y1$.

• Experiment with window settings until you find settings that show both turning points. The graph in Figure 29 shows one example.

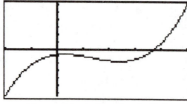

Figure 29. A Graph of $f(x)$

Step 2: Approximate the coordinates of the turning point associated with the local maximum (the y-coordinate of the peak on the graph) as follows:

- In the GRAPH menu, press $\boxed{\text{MORE}}$ $\boxed{\text{F1}}$ to access the GRAPH/MATH menu.

- Specify a small x-interval that contains the x-coordinate associated with the local maximum of the function. Here's how:

On the TI-86:

Press $\boxed{\text{F5}}$ for FMAX.

Use the right or left arrow keys to move the cursor (a box with a blinking $\times$) to the left of the turning point associated with the local maximum (the peak).

Press $\boxed{\text{ENTER}}$ to mark with a black triangle the left bound of this interval.

Use the right arrow key to move the cursor slightly to the right of this turning point.

Press $\boxed{\text{ENTER}}$ to mark with a black triangle the right bound of this interval.

On the TI-85:

Press $\boxed{\text{F1}}$ for LOWER.

Use the right or left arrow keys to move the cursor (a + sign with a blinking center pixel) to the left of the turning point associated with the local maximum (the peak).

Press $\boxed{\text{ENTER}}$ to mark with a black triangle the left bound of this interval.

Press $\boxed{\text{F2}}$ for UPPER.

Use the right arrow key to move the cursor slightly to the right of this turning point.

Press $\boxed{\text{ENTER}}$ to mark with a black triangle the right bound of this interval.

Press $\boxed{\text{MORE}}$ $\boxed{\text{F2}}$ for FMAX.

- Your calculator needs a guess for the turning point. Use the arrow keys to position the cursor on the turning point where the graph peaks. (See Figure 30.) Then press $\boxed{\text{ENTER}}$.

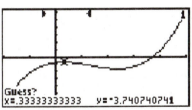

Figure 30. Guess for Local Maximum

- Read off the coordinates of this turning point at the bottom of your screen. The local maximum (y-coordinate of this turning point) should be approximately -3.73.

Step 3: Approximate the coordinates of the turning point associated with the local minimum (the y-coordinate of the valley on the graph).

Press $\boxed{\text{GRAPH}}$ $\boxed{\text{MORE}}$ $\boxed{\text{F1}}$ to access the GRAPH/MATH menu. Adapt the instructions for Step 2 using FMIN to find the coordinates of the turning point associated with the local minimum (the valley). You should get a value for y that is close to -8.42.

Approximating Instantaneous Rates of Change

You can approximate the instantaneous rate of change of a function $f(x)$ at $x = a$ by evaluating the average rate of change of the function over a small interval containing a. For example, you might choose a symmetric interval about a of the form $(a - h, a + h)$ for some small value of h. Using this interval, your approximation of the instantaneous rate of change would be

$$\frac{f(a+h) - f(a-h)}{(a+h) - (a-h)} = \frac{f(a+h) - f(a-h)}{2h}.$$

This is called the central difference quotient.

Example: Use the central difference quotient to approximate the instantaneous rate of change of the function $f(x) = x^3 - 4x^2 + 2x - 4$ at $x = 1$ and $x = 3$. Use $h = .01$.

The formula for the central difference quotient using $a = 1$ and $h = .01$ is $\frac{f(1+.01) - f(1-.01)}{2(.01)}$. You'll do this problem in three steps. In Step 1, you'll enter the function. In Steps 2 and 3, you'll approximate the instantaneous rate of change of $f(x)$ at $x = 1$ and $x = 3$, respectively.

Step 1: Enter $f(x)$ as $y1$ and return to the home screen.
- Press $\boxed{\text{GRAPH}}$ $\boxed{\text{F1}}$. If you completed the previous example, you may already have $f(x)$ stored as $y1$. If not, erase any previously stored functions and then enter $f(x)$ as $y1$.

- Press $\boxed{\text{2nd}}$ $\boxed{\text{QUIT}}$ to return to the home screen.

Step 2: Approximate the instantaneous rate of change of $f(x)$ at $x = 1$.
Enter the formula for the central difference quotient using $x = 1$ and $h = .01$. Make sure you enter the entire formula before pressing enter. Here are the keystrokes.

On the TI-86:
- Press $\boxed{(}$ $\boxed{\text{2nd}}$ $\boxed{\text{CATLG-VARS}}$ $\boxed{\text{MORE}}$ $\boxed{\text{F4}}$ for EQU. Use the down arrow key to highlight $y1$ and then press $\boxed{\text{ENTER}}$.

- Press $\boxed{(}$ $\boxed{1}$ $\boxed{+}$ $\boxed{.}$ $\boxed{0}$ $\boxed{1}$ $\boxed{)}$ $\boxed{-}$.

- Press $\boxed{(}$ $\boxed{\text{2nd}}$ $\boxed{\text{CATLG-VARS}}$ $\boxed{\text{MORE}}$ $\boxed{\text{F4}}$ for EQU. Use the down arrow key to highlight $y1$ and then press $\boxed{\text{ENTER}}$.

- Press $\boxed{(}$ $\boxed{1}$ $\boxed{-}$ $\boxed{.}$ $\boxed{0}$ $\boxed{1}$ $\boxed{)}$ $\boxed{)}$.

- Press $\boxed{\div}$ $\boxed{(}$ $\boxed{2}$ $\boxed{\times}$ $\boxed{.}$ $\boxed{0}$ $\boxed{1}$ $\boxed{)}$

- Press $\boxed{\text{ENTER}}$. Your screen should match Figure 31.

On the TI-85:

You will compute the values of $f(1 + .01)$ and $f(1 - .01)$ and store them for use in the central difference quotient.

- Store $f(1 + .01)$ as A:

  Press $\boxed{1}$ $\boxed{+}$ $\boxed{.}$ $\boxed{0}$ $\boxed{1}$ $\boxed{\text{STO} \triangleright}$ $\boxed{x\text{-VAR}}$

  Press $\boxed{\text{2nd}}$ $\boxed{:}$ (same key as $\boxed{.}$).

  Press $\boxed{\text{2nd}}$ $\boxed{\text{VARS}}$ $\boxed{\text{MORE}}$ $\boxed{\text{F3}}$ for EQU.

  Move the triangle so that it points to $y1$ and press $\boxed{\text{ENTER}}$.

  Press $\boxed{\text{STO} \triangleright}$ $\boxed{A}$ (same key as $\boxed{\text{LOG}}$).

  Press $\boxed{\text{ENTER}}$.

- Store $f(1 - .01)$ as B:

  Press $\boxed{\text{2nd}}$ $\boxed{\text{ENTRY}}$.

  Change the plus to a minus and the A to B.

  Press $\boxed{\text{ENTER}}$.

- Enter the formula for the central difference quotient:

  Press $\boxed{(}$ $\boxed{\text{ALPHA}}$ $\boxed{A}$ $\boxed{-}$ $\boxed{\text{ALPHA}}$ $\boxed{B}$ $\boxed{)}$

  Press $\boxed{\div}$ $\boxed{(}$ $\boxed{2}$ $\boxed{\times}$ $\boxed{.}$ $\boxed{0}$ $\boxed{1}$ $\boxed{)}$

  Press $\boxed{\text{ENTER}}$. Your screen should match Figure 31.

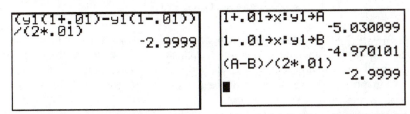

Figure 31. Central Difference Quotient on the TI-86 (Left) and TI-85 (Right)

Step 3: Approximate the instantaneous rate of change of $f(x)$ at $x = 3$.

On the TI-86:

Press $\boxed{\text{2nd}}$ $\boxed{\text{ENTRY}}$ and edit your work from the previous step. Change the two x-values of 1 to 3. Press $\boxed{\text{ENTER}}$. The result should be approximately 5.0.

On the TI-85:

Repeat Step 2 this time storing $f(3 + .01)$ in A and $f(3 - .01)$ in B.

In this example, you approximated instantaneous rates of change directly from a formula. You might prefer using your calculator's **nDer** command instead. For more information, check the topic *Approximating Instantaneous Rates of Change* in Chapter 4 (pages 79 and 80).

Graphing a Rational Function

Warning! *If the numerator or the denominator of a rational function consists of more than one term, you must enclose it in parentheses when you enter it into your calculator.*

Example: Graph $r(x) = \dfrac{x^2 - 1}{x - 3}$ in the standard viewing window. Then adjust window settings to display the key graphical features of this function.

- Press $\boxed{\text{GRAPH}}$ $\boxed{\text{F1}}$ and erase any previously stored functions.

- Enter $r(x)$ as $y1$: press $\boxed{(}$ $\boxed{\text{F1}}$ $\boxed{x^2}$ $\boxed{-}$ $\boxed{1}$ $\boxed{)}$ $\boxed{\div}$ $\boxed{(}$ $\boxed{\text{F1}}$ $\boxed{-}$ $\boxed{3}$ $\boxed{)}$.

- Press $\boxed{\text{EXIT}}$ to return to the GRAPH window. Then press $\boxed{\text{F3}}$ $\boxed{\text{F4}}$ to view the graph in the standard window. Press $\boxed{\text{CLEAR}}$ to remove the menu bar from your screen. Your graph should look similar to the one in Figure 32.

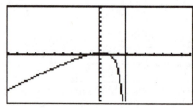

Figure 32. Graph of $r(x)$

The vertical line in the graph shown in Figure 32 indicates that this function has a vertical asymptote at $x = 3$. Remember, this line is not part of the graph of the function. Furthermore, because the domain of this function is all real numbers except $x = 3$, there is a branch of this function's graph that lies to the right of the line $x = 3$. In order to observe this branch, you will have to adjust the window setting for yMax.

- Press $\boxed{\text{GRAPH}}$ to return to the GRAPH menu and the $\boxed{\text{F2}}$ to access the window settings. Change the setting for yMax to 20. Then press $\boxed{\text{F5}}$. You should now see both branches of the graph along with a nearly vertical line that is not part of the graph of this function.

- To remove the vertical line, change the format from DRAWLINE to DRAWDOT. Here's how:
  In the GRAPH menu, press $\boxed{\text{MORE}}$ $\boxed{\text{F3}}$ for FORMT.
  Change the setting from DRAWLINE to DRAWDOT.
  Then press $\boxed{\text{F5}}$. The nearly vertical line should now be gone.

- Return the format setting to DRAWLINE for the next Example.

Do <u>not</u> clear this function from your calculator's memory until after you have completed the next topic, *Zooming Out.*

Zooming Out

Example: Graph the function $r(x) = \dfrac{x^2 - 1}{x - 3}$ in the standard viewing window and then zoom out by a factor of four several times.

For this example, observe what happens to the appearance of the graph of $r(x)$ as we "back away" by increasing the width and height of the viewing window. From the previous example, you should already have the formula for $r(x)$ stored as $y1$.

- In the GRAPH menu, press $\boxed{F3}$ $\boxed{F4}$ to graph the function in the standard viewing window. In the GRAPH menu, press $\boxed{F2}$ to access the window settings. Set xScl and yScl equal to 0. (This turns off the tick marks that appear on the axes. If you skip this step, the axis will get crowded with tick marks when you zoom out.) Then press $\boxed{F5}$ to return to the graph.

- Check the settings for the zoom factor. From the GRAPH menu, press $\boxed{F3}$ $\boxed{MORE}$ $\boxed{MORE}$. Then press the F-key corresponding to ZFACT (on the TI-86, press $\boxed{F2}$; on the TI-85, press $\boxed{F1}$). Your screen should match the one in Figure 33. (If it doesn't, change the settings for xFact and yFact to 4.) Press $\boxed{F5}$ to return to the graph.

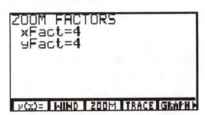

Figure 33. Settings for Zoom Factors

- Press $\boxed{F3}$ $\boxed{F3}$ for ZOUT. A blinking pixel marks the center of the zoom. (If you wish to change the zoom center, use the arrow keys to move the cursor to a new location.) Press $\boxed{ENTER}$ to view the graph over wider x- and y-intervals.

- Press $\boxed{ENTER}$ again to zoom out a second time. (Compare your zoomed out graphs with those in Figure 34.)

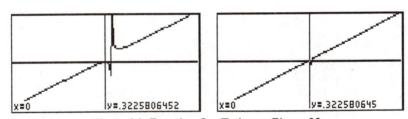

Figure 34. Zooming Out Twice on Figure 32

- Now, press $\boxed{GRAPH}$ $\boxed{F2}$ to observe the effect on the window settings of zooming out twice.

The default setting for ZoomOut widens both the x- and y-intervals by a factor of 4 each time that it is applied. In the previous example, you zoomed out twice. Therefore, the x- and y-intervals are 16 times wider than they were before you zoomed out.

In the next example, you'll change the default zoom settings in order to observe a graph of a function that begins to act like its horizontal asymptote.

Example: Graph $q(x) = \dfrac{5x^2 + 20x - 105}{2x^2 + 2x - 60}$ in the standard viewing window.

- Press $\boxed{\text{GRAPH}}$ $\boxed{\text{F1}}$, erase any stored functions, and then enter $q(x)$ as $y1$. Press $\boxed{\text{EXIT}}$ $\boxed{\text{F3}}$ $\boxed{\text{F4}}$. If you have entered the function correctly, your graph should look like the one in Figure 35.

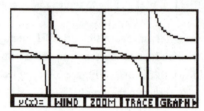

Figure 35. Graph of $q(x)$

Based on this graph, you may suspect that the function $q(x)$ has a horizontal asymptote but this is not at all obvious. Let's observe the function's graph over increasingly wide x-intervals to see if it begins to behave like a horizontal line. Instructions on how to zoom out in the horizontal direction follow.

- In the GRAPH menu, press $\boxed{\text{F2}}$ and set xScl to 0. Then press $\boxed{\text{F5}}$ to return to the graph.

- Press $\boxed{\text{F3}}$ $\boxed{\text{MORE}}$ $\boxed{\text{MORE}}$. Then press the F-key corresponding to ZFACT. Leave the setting for xFact at 4; change the setting for yFact from 4 to 1. Press $\boxed{\text{F5}}$ to return to the graph.

- Press $\boxed{\text{F3}}$ $\boxed{\text{F3}}$ for ZOUT. Then press $\boxed{\text{ENTER}}$ to widen the x-interval by a factor of four. Press $\boxed{\text{ENTER}}$ several more times to continue widening the x-interval. Your graph should begin to resemble its horizontal asymptote $y = 2.5$.

CHAPTER 6

Graphing Trigonometric Functions

Three of the six basic trigonometric functions are built-in functions on the TI-85/86: sine, ⟦SIN⟧, cosine ⟦COS⟧, and tangent ⟦TAN⟧. Before graphing any of these functions, you should first check that your calculator is set to radian mode: press ⟦2nd⟧ ⟦MODE⟧ and check to see that Radian is highlighted. (If Degree is highlighted, use the arrow keys to position the cursor over Radian and press ⟦ENTER⟧.) Press ⟦2nd⟧ ⟦QUIT⟧ to return to the home screen.

Example: Graph $y = \sin(x)$ and $y = \csc(x)$ in the trigonometric viewing window.
- Press ⟦GRAPH⟧ ⟦F1⟧ and erase any stored functions. Then enter the function $\sin(x)$ as $y1$ by pressing ⟦SIN⟧ ⟦(⟧ ⟦F1⟧ ⟦)⟧ ⟦ENTER⟧.

  Warning! *Since $\csc(x)$ is defined as $\frac{1}{\sin(x)}$, you will raise sin(x) to the -1 power using the reciprocal key* ⟦x^{-1}⟧. *You could also divide 1 by sin(x). However, you* <u>cannot</u> *use the* ⟦SIN^{-1}⟧ *key which is reserved for the inverse sine function.*

- Enter $\csc(x)$ as $y2$: Press ⟦(⟧ ⟦SIN⟧ ⟦(⟧ ⟦F1⟧ ⟦)⟧ ⟦)⟧ ⟦2nd⟧ ⟦x^{-1}⟧ (same key as ⟦EE⟧).

- To graph these functions in the trig viewing window, press ⟦EXIT⟧ ⟦F3⟧ ⟦MORE⟧ ⟦F3⟧ for ZTRIG. Press ⟦CLEAR⟧ to remove the menu bar from the bottom of your screen. Your graph should be similar to the one in Figure 36.

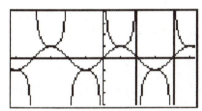

Figure 36. Graphs of $\sin(x)$ and $\csc(x)$

- Press ⟦GRAPH⟧ ⟦F2⟧ to observe the settings for the trig viewing window.

The trigonometric viewing window gives a good picture of the graphs of $\sin(x)$ and $\csc(x)$. Keep in mind, however, that it is not the best window for viewing all trigonometric functions. For example, it would not be a good viewing window for the function $y = 5\cos(10x)$. For this function, you would need to adjust the window settings in order to display the key features of its graph.

Approximating Instantaneous Rates of Change

You can approximate the instantaneous rate of change of a function $f(x)$ at $x = a$ by evaluating the average rate of change of the function over a small interval containing a. For example, you might choose a

89

symmetric interval about a of the form $(a - h, a + h)$ for some small value of h. Using this interval, your approximation of the instantaneous rate of change would be

$$\frac{f(a + h) - f(a - h)}{(a + h) - (a - h)} = \frac{f(a + h) - f(a - h)}{2h}.$$

Example: Use the previous formula (called the symmetric difference quotient) with $h = .01$ to approximate the instantaneous rate of change of $f(x) = \cos(x)$ at $x = \frac{\pi}{2}$ and $x = \pi$.

On the home screen, you'll enter the formula for the central difference quotient. The key strokes follow. Notice that both numerator and denominator are enclosed in parentheses. The formula is quite lengthy. Don't press ENTER until you reach the end.

- On the home screen, enter the numerator for the first problem:
  Press (COS (2nd π ÷ 2 + . 0 1).
  Press − COS (2nd π ÷ 2 − . 0 1)).

- Press ÷ .

- Enter the denominator: press (2 × . 0 1)

- Press ENTER . You should get approximately − 1.0.

- Press 2nd ENTRY so you can edit the previous problem.

- Edit the entry by changing $\pi/2$ to π. Then press ENTER . You should get 0. (Can you figure out why?)

Restricting the Domain of a Function

Only one-to-one functions have inverses. In order to define the inverse of a trigonometric function, you must restrict its domain so that the restricted trigonometric function is one-to-one. In the next example, you'll do that for the sine function.

Example: Graph $f(x) = \sin(x)$ on the restricted domain $-\frac{\pi}{2} \leq x \leq \frac{\pi}{2}$. Use a $[-2,2] \times [-2,2]$ window.

Note: Your calculator can only interpret one inequality at a time. Hence it cannot interpret the condition $-\frac{\pi}{2} \leq x \leq \frac{\pi}{2}$ written in this form. Instead, you will need to enter the equivalent condition $-\frac{\pi}{2} \leq x$ and $x \leq \frac{\pi}{2}$.

- Press GRAPH F1 and clear any previously stored functions.

- Enter the sine function as $y1$: press SIN (F1). Don't press enter just yet.

- Add the restriction on the domain as follows:
  Press × ((-) 2nd π ÷ 2 2nd TEST F4 for $\leq$. Then press x-VAR .

Press $\boxed{\text{2nd}}$ $\boxed{\text{BASE}}$ $\boxed{\text{F4}}$ for BOOL. Then press $\boxed{\text{F1}}$ for *and*.
Press $\boxed{x\text{-VAR}}$ $\boxed{\text{2nd}}$ $\boxed{\text{TEST}}$ $\boxed{\text{F4}}$ for $\leq$. Then press $\boxed{\text{2nd}}$ $\boxed{\pi}$ $\boxed{\div}$ $\boxed{2}$ $\boxed{)}$.

At the end of these key-strokes, you should have $y1= \sin(x)*(-\pi/2 \leq x$ and $x \leq \pi/2)$. You will need to use the left and right arrow keys to see the entire function.

- Press $\boxed{\text{EXIT}}$ twice to return to the GRAPH menu and then press $\boxed{\text{F2}}$ and adjust the settings for a $[-2,2] \times [-2,2]$ window. Press $\boxed{\text{F5}}$ to view the graph.

- Notice that nearly vertical line segments connect the ends of the graph to the x-axis. To remove these line segments, you will need to change the format to DRAWDOT. From the GRAPH menu, press $\boxed{\text{MORE}}$ $\boxed{\text{F3}}$ for FORMT and change the settings from DRAWLINE to DRAWDOT. Then press $\boxed{\text{F5}}$.

Notice that $f(x) = \sin(x)$ restricted to the domain $-\frac{\pi}{2} \leq x \leq \frac{\pi}{2}$ is one-to-one. Hence it has an inverse.

Do <u>not</u> clear this function from your calculator's memory until after you have completed the next example.

Example: Graph the restricted sine function from the previous example and its inverse on the same screen. Use a $[-2,2] \times [-2,2]$ window.
The function $f(x)$, the restricted sine function from the previous example, should already be stored as $y1$.

- Press $\boxed{\text{GRAPH}}$ $\boxed{\text{F1}}$. Press $\boxed{\text{ENTER}}$ to position the cursor opposite $y2$.

- Press $\boxed{\text{2nd}}$ $\boxed{\text{SIN}^{-1}}$ $\boxed{(}$ $\boxed{\text{F1}}$ $\boxed{)}$.

- Press $\boxed{\text{EXIT}}$ to return to the GRAPH menu. Press $\boxed{\text{F2}}$ and check that the window settings are for a $[-2,2] \times [-2,2]$ window. Then press $\boxed{\text{F5}}$ for GRAPH. Your screen should match Figure 37.

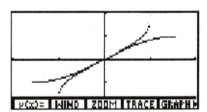

Figure 37. Graphs $\sin(x)$ and $\sin^{-1}(x)$

- Return the format to DRAWLINE before beginning the next example.

91

Finding Values of Inverse Trigonometric Functions

Example: Approximate the values of $\cos^{-1}(0.8)$ and $\cos^{-1}(1.8)$.

- To approximate $\cos^{-1}(0.8)$, from the home screen press 2nd $\cos^{-1}$ (. 8) ENTER . Your answer should be approximately .644.

- To compute $\cos^{-1}(1.8)$, press 2nd ENTRY , position the cursor over the decimal point, press 2nd INS 1 ENTER . On many calculator's, you would get an error message. However, the TI-85/86 calculates this value as a complex number and reports its value as approximately (0,1.193) or $1.193i$.

Example: Find $\tan^{-1}(\frac{1}{2})$.

Press 2nd TAN^{-1} (1 ÷ 2) ENTER . You should get approximately .464.

Solving Equations Involving Trigonometric Functions

You can approximate solutions to equations involving trigonometric functions graphically.

Example: Approximate the solution to the equation $\cos(x) = 2\sin(x)$.

To approximate the solutions to this equation, first graph $y = \cos(x)$ and $y = 2\sin(x)$. Then approximate their points of intersection.

- Press GRAPH F1 and clear any previously stored functions. Then enter $\cos(x)$ as $y1$ and $2\sin(x)$ as $y2$.

- Press EXIT F3 MORE F3 to view the graphs. There are five intersection points, which correspond to five solutions, visible in this window. (See Figure 38.) If you widen the x-interval, you will see more intersection points, which correspond to additional solutions.

- Approximate the solution, the x-coordinate of the intersection point, that is closest to zero.
  From the GRAPH menu, press MORE F1 to access the GRAPH/MATH menu.
  Press MORE and then press the F-key corresponding to ISECT (press F3 on a TI-86; press F5 on a TI-85).

- Press ENTER to designate $y = \cos(x)$ as the first curve. On the TI-85, proceed to the next bullet because the cursor automatically jumps to the second curve. On the TI-86, press ENTER a second time to designate $y = 2\sin(x)$ as the second curve.

- Use the right arrow key to move the cursor over the intersection point with x-value closest to zero. Your screen should be similar to the one in Figure 38. Then press ENTER and read the x-value from the bottom of the screen. You should get approximately 0.463.

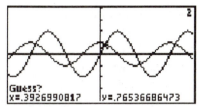

Figure 38. Guessing the Point of Intersection

- Next, press ⎡GRAPH⎤. Then repeat the steps outlined in the previous three bullets to approximate the other solution that lies in the interval $0 < x < 2\pi$. You should get approximately 3.605.

Because the functions $\cos(x)$ and $2\sin(x)$ have period 2π, you can obtain other approximate solutions to $\cos(x) = 2\sin(x)$ by adding integer multiples of 2π to the two solutions that you've just found.

In the next example, you'll use solver to find the solutions to $\cos(x) = 2\sin(x)$.

Example: Use SOLVER to find the solutions to $\cos(x) = 2\sin(x)$ in the interval between 0 and 2π.
You should already have $\cos(x)$ and $2\sin(x)$ stored as $y1$ and $y2$, respectively. If not, enter these functions. First, you'll find an approximate solution in the interval from 0 to $\frac{\pi}{2}$ and then one in the interval from $\frac{\pi}{2}$ to 2π.

- From the home screen, press ⎡2nd⎤ ⎡SOLVER⎤ (same key as ⎡GRAPH⎤).

- Enter the equation $y1 = y2$ opposite equ:
   Press the F-key corresponding to $y1$.
   Press ⎡ALPHA⎤ ⎡=⎤ (same key as ⎡STO ▷⎤)
   Press the F-key corresponding to $y2$.

   (If you prefer, you could enter the equation $\cos(x) = 2\sin(x)$ instead.)

- Press the down arrow key to activate the equation. A solution in the interval specified by *bound* = will appear. (This may or may not be the solution that you want. Proceed as if it's not.)

- Find the solution between 0 and $\frac{\pi}{2}$:
   The cursor should be opposite $x =$. Press ⎡CLEAR⎤.
   Press the down arrow key to move the cursor opposite *bound* =. Press ⎡CLEAR⎤.
   Enter the interval as a lower bound and an upper bound written in list form: press ⎡2nd⎤ ⎡LIST⎤
      ⎡F1⎤ for {. Then press ⎡0⎤ ⎡,⎤ ⎡2nd⎤ ⎡π⎤ ⎡÷⎤ ⎡2⎤ ⎡F2⎤ for }.
   Press ⎡EXIT⎤ to return to the SOLVER screen.
   Press the up arrow key to move the cursor opposite $x =$ and then press ⎡F5⎤ for SOLVE. You
      should get approximately 0.0464.

- Find the solution between $\frac{\pi}{2}$ and 2π:
   Clear the entries opposite $x =$ and *bound* = .

Move the cursor opposite *bound =* and enter the interval.

Adapt the instruction in the previous bullet to find the solution. You answer should be approximately 3.605.

Problems Inherent in the Technology
(Don't Believe Everything That You See!)

Your viewing screen consists of a grid of pixels. When a pixel is *on*, it shows up as a dark square dot on the screen. Graphs are formed by turning on a series of pixels. This method of producing graphs can sometimes produce misleading images.

Example: Graph $g(x) = \sin(x)$ over increasingly wide x-intervals.
- Press GRAPH F1 and clear any previously stored functions.

- Now, begin by entering $\sin(x)$ as $y1$. Then press EXIT to return to the GRAPH menu. Press F3 MORE F3 to graph $g(x)$ in the trigonometric viewing window.

- Turn off the tick marks for the x-axis: press F2 and set xScl = 0.

- Change the zoom factors. From the GRAPH menu, press F3 MORE MORE. Press the F-key corresponding to ZFACT. Set xFact = 10 and yFact = 1 Then press F5.

Get ready to have some fun. When $\sin(x)$ is graphed in the trig viewing window, how many complete sine waves do you see? Each time you increase the width of the x-interval by a factor of 10, you should see 10 times as many cycles of the sine wave.

- From the GRAPH menu, press F3 F3 for ZOUT. Then press ENTER. Do you see 10 times as many cycles? Press ENTER to zoom out again. Do you see 10 times as many cycles as in the previous graph?

The graph produced by the second zoom out shows fewer cycles than the previous graph. In producing the graphs, your calculator does not have enough pixels to capture all the oscillations that are part of the actual graph. In this case, the small subset of points from the actual graph that your calculator chooses to represent with darkened pixels presents a very misleading picture of the features of the actual graph.

- Press ENTER several more times to view other interesting graphs, all of which give misleading pictures of the function's key graphical features.

CHAPTER 7

Finding Values of Trigonometric Functions

With the TI-85/86 you can solve problems in right-triangle trigonometry. To compute the sine, cosine, or tangent of an angle measured in degrees, you will first need to change the mode setting to Degree. Press [2nd] [MODE]. Use the arrow keys to position the cursor on Degree, and press [ENTER] to save the setting. Press [2nd] [QUIT] to return to the home screen.

Example: Compute sin(60°) and sin(29°).
- Press [SIN] [(] [6] [0] [)] [ENTER]. You should get approximately 0.866.

- Edit the previous command to compute sin(29°): press [2nd] [ENTRY], change 60 to 29, and press [ENTER].

Example: Compute tan(90°).

Press [TAN] [(] [9] [0] [)] [ENTER]. Your calculator will respond with the error message: ERR: DOMAIN because the tangent function is undefined at 90°. Press [F5] to quit.

Finding Values of Inverse Trigonometric Functions

Example: Approximate the value of sin⁻¹(.3).
Press [2nd] [SIN⁻¹] [(] [.] [3] [)] [ENTER]. You should get approximately 17.5°. (Because you are in degree mode, the output will be in degrees.)

Example: Approximate the value of tan⁻¹(10).
Press [2nd] [TAN⁻¹] [(] [1] [0] [)] [ENTER]. You should get approximately 84.3°.

CHAPTER 8

Working With Parametric Equations

Before you can use your calculator to graph parametric equations, you'll need to change your calculator from function mode to parametric mode. Press $\boxed{\text{2nd}}$ $\boxed{\text{MODE}}$ and adjust your mode settings to match those in Figure 39. Then press $\boxed{\text{2nd}}$ $\boxed{\text{QUIT}}$ to return to the home screen.

Figure 39. Mode Settings

Example: Graph the parametric equations $x(t) = 10 - 3t$ and $y(t) = 1 + 4t$ in the standard viewing window.

- Press $\boxed{\text{GRAPH}}$ $\boxed{\text{F1}}$ for $E(t)$. Notice in parametric mode, you enter pairs of formulas, one for $x(t)$, and the other for $y(t)$. If you have any parametric equations stored in memory, erase them by positioning the cursor on each equation and pressing $\boxed{\text{CLEAR}}$.

- Enter the formula for $x(t)$ as $xt1$: press $\boxed{1}$ $\boxed{0}$ $\boxed{-}$ $\boxed{3}$ $\boxed{\text{F1}}$ $\boxed{\text{ENTER}}$.

- Enter the formula for $y(t)$ as $yt1$: press $\boxed{1}$ $\boxed{+}$ $\boxed{4}$ $\boxed{\text{F1}}$.

- Press $\boxed{\text{EXIT}}$ $\boxed{\text{F3}}$ $\boxed{\text{F4}}$ to graph this set of parametric equations in the standard viewing window.

- Press $\boxed{\text{F2}}$ to observe the settings for the standard window. Your screen should match the one in Figure 40. (If tMax = 360 instead of approximately 6.28, change your mode settings to Radian.)

```
WINDOW
 tMin=0
 tMax=6.28318530718
 tStep=.130899693899...
 xMin=-10
 xMax=10
↓xScl=1
E(t)=  WIND  ZOOM  TRACE  GRAPH▶
```

Figure 40. Standard Window Settings

Notice that the settings for the standard viewing window include settings for t: tMin = 0, tMax $\approx 2\pi$, and tStep $\approx \frac{\pi}{24}$. The settings for the x-interval and y-interval are the same as in function mode.

- Press $\boxed{\text{F5}}$ to return to the graph and then press $\boxed{\text{F4}}$ for TRACE. Notice that the trace begins at the point corresponding to $t = t\text{Min}$, or in this case $t = 0$. Press the right arrow to move along the graph in t-increments of approximately 0.13.

Do <u>not</u> clear these parametric functions until you have completed the next example.

Example: Graph the position of a dot as it moves along the path $x(t) = 10 - 3t$ and $y(t) = 1 + 4t$ at one-second intervals from time $t = 0$ seconds to $t = 10$ seconds.
You should already have the parametric equations stored in your calculator from your work on the previous example.

- In the GRAPH menu, press $\boxed{\text{MORE}}$ $\boxed{\text{F3}}$ for FORMT and change the setting from DRAWLINE to DRAWDOT. Press $\boxed{\text{F5}}$ to return to the graph.

- Press $\boxed{\text{F2}}$. Adjust the parameter settings for t: $t\text{Min} = 0$, $t\text{Max} = 10$, $t\text{Step} = 1$.

- In the GRAPH menu, press $\boxed{\text{F3}}$ $\boxed{\text{MORE}}$ $\boxed{\text{F1}}$ for ZFIT. You should see a group of isolated dots that fall on a line similar to those shown in Figure 41.)

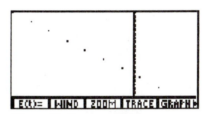

Figure 41. Dots on a Line

- Press $\boxed{\text{F4}}$ for TRACE. Then press the right arrow key repeatedly. Watch the cursor move from one dot to the next.

Return your calculator to DRAWLINE format before beginning the next topic.

Using Square Scaling in Parametric Mode

Example: Graph the circle described by the set of parametric equations $x(t) = 5\cos(t)$, $y(t) = 5\sin(t)$ first in the standard viewing window and then in a square-scaling window.
- Press $\boxed{\text{GRAPH}}$ $\boxed{\text{F1}}$ and clear any stored equations. Then enter the formula for $x(t)$ as $xt1$ and the formula for $y(t)$ as $yt1$. Press $\boxed{\text{EXIT}}$ to return to the GRAPH menu.

- Press $\boxed{\text{F3}}$ $\boxed{\text{F4}}$. Your graph should look more elliptical than circular.

- To change to square scaling, press $\boxed{\text{F3}}$ $\boxed{\text{MORE}}$ $\boxed{\text{F2}}$ for ZSQR. Press $\boxed{\text{CLEAR}}$ to remove the menu bar. Your graph should now look like a circle.

Combining Two Sets of Parametric Equations

In Lab 8B, Bézier Curves, you are asked to form a new set of parametric equations from a combination of two other sets of parametric equations.

Example: Suppose that you have two sets of parametric equations

$$S1: \quad x_1 = 2t + 1 \qquad S_2: \quad x_2 = t - 5$$
$$y_1 = -3t + 5 \qquad\qquad y_2 = 4t - 3$$

and that you want to graph the combination $(1 - t)S_1 + tS_2$ over the interval $0 \leq t \leq 1$.

You'll tackle this problem in three steps. In Step 1, you'll enter the equations for S_1 and S_2. In Step 2, you'll form the combination. In Step 3, you'll graph the combination.

Step 1: Enter the parametric equations S_1 and S_2.
- Be sure that your calculator is in parametric mode. Press GRAPH   F1 and clear any stored functions.

- Enter the equations for S_1 as $xt1$ and $yt1$ and the equations for S_2 as $xt2$ and $yt2$.

Step 2: Enter the x- and y-equations for the combination $(1 - t)S_1 + tS_2$:

$$x_3 = (1 - t)x_1 + tx_2$$
$$y_3 = (1 - t)y_1 + ty_2$$

- The cursor should be opposite $xt3$. Enter the equation for x_3:
  Press (1   −   F1   )   × .
  Press F2   1 for $xt1$.
  Press +   F1   × .
  Press F2   2 for $xt2$.
  Press ENTER .

- The cursor should now be opposite $yt3$. Enter the equation for y_3.
  Press (1   −   F1   )   ×
  Press F3   1 for $yt1$.
  Press +   F1   × .
  Press F3   2 for $yt2$.
  Press ENTER .

Step 3: Graph the combination $(1 - t)S_1 + tS_2$. Here's how:
- Unselect (turn off) parametric equations $xt1$, $yt1$, $xt2$, and $yt2$. To do this, move the cursor to the line containing $xt1$. Press F5 . Notice that both $xt1$ and $yt1$ have been turned off. Repeat this process to turn off $xt2$ and $yt2$.

• Press ⬚EXIT⬚ to return to the GRAPH menu. Then press ⬚F2⬚. Set tMin = 0, tMax = 1, and tStep = 0.1. Adjust the remainder of the settings for a $[-5,2] \times [-4,6]$ window.

• In the GRAPH menu, press ⬚F5⬚ to graph the combination. Your graph should match the one in Figure 42.

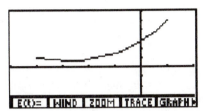

Figure 42. Graphing a Combination

You've reached the end of this guide. Your calculator has many more features in addition to what you've learned from this guide. Consult the manual that came with your calculator to learn more about your calculator's capabilities.

Graphing Calculator Guidebook
TI-89/92Plus/Voyage 200
To Accompany
Precalculus: Concepts in Context, 2e

This guide provides background on the TI-89/92Plus/Voyage 200 graphing calculator that will be useful for *Precalculus: Concepts in Context, 2e*. It consists of a basic tutorial followed by additional instructions relevant to each chapter of your text. Consult your calculator manual for additional calculator features.

BASIC TUTORIAL

0. *The Keyboard*

The three major sections of the keyboard can be classified as follows:

- function keys (below the screen on the TI-89/Voyage 200; to the left of the screen on the TI-92Plus)
- letter keys (The letter keys on the TI-89 are accessed by pressing alpha followed by one of the keys in the last five rows. The TI92Plus/Voyage 200 has a QWERTY keyboard that lies below the screen.)
- calculator-type keys (on the bottom center of the TI-89 and on the bottom right of the TI-92Plus/Voyage 200.)

In addition, you should familiarize yourself with the locations of the keys that control the cursor. These arrow keys are located on the right-hand side of the calculator and point out in four directions as shown in Figure 1.

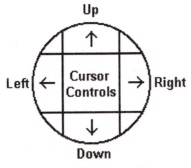

Figure 1. The Arrow Keys

Finally, take a moment to find the ESC and ENTER keys. (Notice that the TI-92Plus/Voyage 200 has several ENTER keys.)

The multiple function keys operate as follows:

- To access a function written in yellow lettering on the TI-89/TI-92Plus or written in blue lettering on the TI-Voyage 200, press the ⎡2nd⎤ key.
  Example: ⎡2nd⎤ ⎡QUIT⎤ (same key as ⎡ESC⎤).
- To access a function written in green, press the ⎡◇⎤ key.
  Example: ⎡◇⎤ ⎡Y=⎤ (on the TI-89, same key as ⎡F1⎤; on the TI-92Plus/Voyage 200, same key as ⎡W⎤).
- On the TI-89, to access letters written in purple, press the ⎡alpha⎤ key.
  Example: ⎡alpha⎤ ⎡A⎤ (same key as ⎡=⎤).

1. *On, Off, and Contrast*

Turn the calculator on by pressing ⎡ON⎤.

You may need to adjust the contrast. Hold down the ⎡◇⎤ key, and then press ⎡+⎤ to darken or ⎡−⎤ to lighten.

To turn your calculator off, press ⎡2nd⎤ ⎡OFF⎤. If you forget, the calculator will automatically turn off after a period of non-use.

2. *Basic Calculations*

When you turn on a TI-89/92Plus, you will see the home screen. You can access the home screen from the TI-Voyage 200 by pressing ⎡◇⎤ ⎡CALC HOME⎤ (same key as ⎡Q⎤).

First, you need to understand something about the sections of your calculator's screen. At the top of the screen is a toolbar. The home screen lies below the toolbar. The entry line (or command line) and status line appear as the last two lines on the screen. See Figure 2 for parts of the screen. (The screen images on all three calculators, TI-89, TI-92, and TI-Voyage 200, are similar. The screens images used throughout this guidebook are from a TI-Voyage 200.)

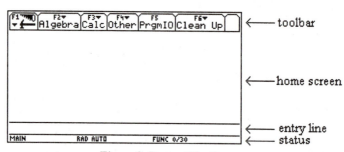

Figure 2. Parts of the Screen

Your calculations will be displayed on the home screen. If, at any time, you want to clear the home screen, press $\boxed{\text{F1}}$ $\boxed{8}$. Clearing the home screen leaves the entry line and status unchanged. You do not need to clear the home screen after each computation. If not cleared, the home screen will keep a history of your work. Previous work can easily be retrieved for use in later calculations.

Work through the following examples to learn about basic computations and mathematical techniques on the TI-89/92Plus/Voyage 200.

Example: Compute 3×4.

After pressing $\boxed{3}$ $\boxed{\times}$ $\boxed{4}$, press $\boxed{\text{ENTER}}$. Note that the original problem, written as $3 * 4$, remains on the entry line. On the home screen, directly above the entry line, the original problem written as $3 \cdot 4$ appears on the left side of the screen and the answer on the right.

The highlighted example, 3*4, will clear automatically when you begin entering the next example. If you prefer to start with a clear entry line, press $\boxed{\text{CLEAR}}$.

Example: Compute $3 + 2 \times 6$ and then $(3 + 2) \times 6$.

After you complete the first calculation, you'll use your calculator's editing features to save keystrokes when entering the second problem.

- First, press $\boxed{3}$ $\boxed{+}$ $\boxed{2}$ $\boxed{\times}$ $\boxed{6}$ $\boxed{\text{ENTER}}$. Do not press $\boxed{\text{CLEAR}}$.

- Complete the second calculation as follows:

Press the left arrow key to move the cursor (a blinking vertical line) to the beginning of the entry line.

Press $\boxed{(}$ (the key directly above $\boxed{7}$) to insert the left parenthesis.

Press the right arrow key to position the cursor after the 2, insert the right parenthesis, $\boxed{)}$ (the key directly above $\boxed{8}$).

Press $\boxed{\text{ENTER}}$. You should get 30.

Example: Compute 8^3.

Press $\boxed{8}$ followed by $\boxed{\wedge}$ $\boxed{3}$ $\boxed{\text{ENTER}}$. Notice the "pretty print" feature of this calculator; it writes 8^3 on the home screen.

Example: Compute $\sqrt{16}$.

Press $\boxed{\text{2nd}}$ $\boxed{[\sqrt{\ }]}$ (same key as times, $\boxed{\times}$) followed by $\boxed{1}$ $\boxed{6}$ $\boxed{\text{ENTER}}$. *OOPS! The calculator automatically inserted a left parenthesis. If you didn't close the parentheses by pressing $\boxed{)}$, you'll get an error message. Press $\boxed{\text{ESC}}$. Go back and insert the right parenthesis and then press* $\boxed{\text{ENTER}}$.

Example: Compute $\sqrt{-16}$.

***Warning!** The TI-89 has two minus keys, $\boxed{-}$ and $\boxed{(-)}$, to differentiate between the operation of subtraction (such as $3 - 2 = 1$) and the opposite of the positive number 16, namely -16.*

Press $\boxed{\text{2nd}}$ $\boxed{[\,\sqrt{\ }\,]}$ followed by $\boxed{(-)}$ (the key to the left of $\boxed{\text{ENTER}}$, last row of the calculator pad) $\boxed{1}$ $\boxed{6}$ $\boxed{)}$ $\boxed{\text{ENTER}}$. *OOPS, again. If your calculator is in Real format, then you will not be allowed to take the square root of a negative number. Press* $\boxed{\text{ESC}}$ *to exit the error message. The last line of the home screen records the type of error. (See Figure 3.)*

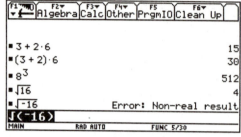 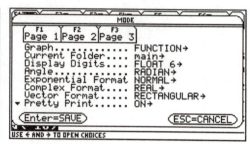

Figure 3. Error Message   Figure 4. Page 1 of the Mode Screen

Press $\boxed{\text{MODE}}$. Your mode screen should be similar to the one in Figure 4. Notice that FUNCTION is highlighted. Press the down arrow to highlight REAL, the setting for Complex Format. Press the right arrow to bring down the menu. Press $\boxed{2}$ to change the format to RECTANGULAR and then press $\boxed{\text{ENTER}}$ to save the setting and return to the home screen.

Now try this problem again. (If the problem is still on the entry line, just press $\boxed{\text{ENTER}}$.) This time the answer should be the imaginary number $4 \cdot i$.

Press $\boxed{\text{MODE}}$ again and return the setting for Complex Format to REAL. (We will be dealing only with real numbers in this course.)

Example: Find a decimal approximation of $\sqrt{5}$.

- Press $\boxed{\text{MODE}}$ $\boxed{\text{F2}}$ to view the second page of the mode screen. If the setting for Exact/Approximate is AUTO, press $\boxed{\text{ENTER}}$ to return to the home screen. Otherwise, press the down arrow until the setting for Exact/Approximate is highlighted. Press the right arrow key to display the possible settings. Press $\boxed{1}$ for AUTO and then $\boxed{\text{ENTER}}$.

- Press $\boxed{\text{2nd}}$ $\boxed{\sqrt{\ }}$ $\boxed{5}$ $\boxed{)}$ $\boxed{\text{ENTER}}$. You should get the exact answer, which is $\sqrt{5}$.

- Press $\boxed{\diamond}$ $\boxed{\approx}$ (same key as $\boxed{\text{ENTER}}$) for a decimal approximation. You should see the result of 2.23607 on the home screen.

3. Correcting an Error and Editing

If you make an error in calculation, you may be able to correct the error without reentering the problem. You can also use the editing feature to modify previous calculations.

104

Example: Make a deliberate error by entering 3 + + 2 and then correct it.

Press $\boxed{3}$ $\boxed{+}$ $\boxed{+}$ $\boxed{2}$ $\boxed{\text{ENTER}}$. To correct the error, press $\boxed{\text{ESC}}$. The cursor will direct you to the error. Erase one of the plus signs by pressing $\boxed{\leftarrow}$ (the backspace key), and then press $\boxed{\text{ENTER}}$. The correct answer to 3 + 2 will appear.

Example: Enter the problem (3 + 2 × 6)/5. Then change it to (3 + 2 × 6)/7.
- Enter the first problem: press $\boxed{(}$ $\boxed{3}$ $\boxed{+}$ $\boxed{2}$ $\boxed{\times}$ $\boxed{6}$ $\boxed{)}$ $\boxed{\div}$ $\boxed{5}$ $\boxed{\text{ENTER}}$.

- Press the right arrow key to remove the highlighting and to position the cursor to the right of the problem in the entry line. Your screen should be similar to the one in Figure 5.

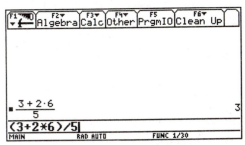

Figure 5. Editing a Calculation

- Press $\boxed{\leftarrow}$ (the backspace key) to erase the 5. Then press $\boxed{7}$ $\boxed{\text{ENTER}}$. If your calculator reports a fraction and you want a decimal approximation, press $\boxed{\diamond}$ $\boxed{\approx}$.

4. Resetting the Memory

After experimenting with settings, you may want to return your calculator's settings to the factory settings. Here's how:

Press $\boxed{\text{2nd}}$ $\boxed{\text{MEM}}$ (same key as $\boxed{6}$) $\boxed{\text{F1}}$.

On the TI-89/Voyage 200, press $\boxed{1}$ for RAM. On the TI-92Plus, skip to the next step.

Press the number for Default (on the TI-89/Voyage 200, press $\boxed{2}$; on the TI-92, press $\boxed{3}$)

Press $\boxed{\text{ENTER}}$ twice.

You may need to adjust the contrast after resetting the calculator. (Instructions for adjusting the contrast appear in Topic 1, page 102.)

5. Graphing

Example: Graph $y = x$, $y = x^2$, and $y = x^3$ in the standard viewing window.

You'll tackle this problem in two steps. In Step 1, you'll enter the function. In Step 2, you'll graph it and then look at the window settings.

Step 1: Enter the three functions as follows.
- Press $\boxed{\diamond}$ $\boxed{\text{Y=}}$ (on the TI-89, same key as F1; on the TI-92Plus/Voyage 200, same key as $\boxed{\text{W}}$).

- If you have functions already stored, highlight a function using the up or down arrow keys, and then press CLEAR to erase the previously stored function.

- Enter the first function to the right of $y1$ by pressing X followed by ENTER. (Note that the function will appear on the entry line until you press ENTER.)

- Enter the second and third functions to the right of $y2$ and $y3$, respectively. (Remember to press ^ to obtain powers.)

Step 2: Set the standard viewing window and graph. Then check the window settings.
- Look at the toolbar at the top of the screen. Press F2 to pull down the Zoom menu. Then press 6 to select ZoomStd. The graphs of all three functions should appear on your screen.

- To check the window settings, press ◇ WINDOW (on the TI-89, same key as F2; on the TI-92Plus/Voyage 200, same key as E). Your screen should match Figure 6.

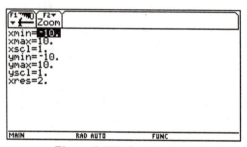

Figure 6. Window Settings

- To exit this menu (or any menu), press 2nd QUIT (same key as ESC).

Example: Graph $y = x$, $y = x^2$, and $y = x^3$, the functions from the previous example, using the window settings shown in Figure 7. (Directions follow on the next page.)

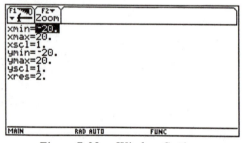

Figure 7. New Window Settings

- Press ◇ WINDOW to access the Window editor. Change the first setting to -20 (*remember to use the* (-) *-key when you enter* -20) and press ENTER. Change the remaining settings to match those in Figure 7.

- Press ◇ GRAPH (on the TI-89, same key as F3; on the TI-92Plus/Voyage 200, same key as R) to view the graphs in the new window.

Example: Zoom in twice on the graph from the previous example to get a magnified view around a point of intersection. Then use Trace to estimate the coordinates of this point.

- Press F2 2 for ZoomIn. Notice that the center of this first zoom (marked by a blinking cursor) will be (0,0). Press ENTER to zoom in on the origin. Your graph should be similar to the one in Figure 8.

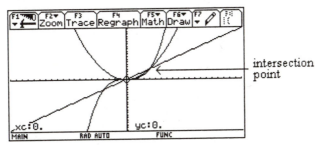

Figure 8. Intersection Point

- Notice the point of intersection to the right of (0,0) (See Figure 8.) Next, you'll zoom in using a center that is close to this point of intersection. Press F2 2 . Use the arrow keys to move the cursor close to this point of intersection. Then press ENTER . The point of intersection should be clearly visible near the center of the screen.

- Press F3 for Trace. Trace along the first curve by pressing the left and right arrow keys. You should see the cursor (a circle with cross hairs) move along the curve. Press the up or down arrow keys to jump from one curve to another.

- Move the cursor directly on top of the intersection point that lies to the right of the origin. Read the approximate values of the x- and y-coordinates from the bottom of the screen. The coordinates should be close to (1,1).

Example: Take a closer look at the function $y = x$ without the other functions. Then view the graph in a window that uses square scaling.

You'll tackle this problem in three steps. In Step 1, you'll unselect the functions $y = x^2$ and $y = x^3$. In Step 2, you'll set up the window and view the graph. In Step 3, you'll clear the stored functions.

Step 1: Remove the graphs of $y = x^2$ and $y = x^3$ from the viewing screen without erasing them as follows:

- Press ◇ Y= to display the stored functions.

- To unselect y2, use the up or down arrow key to highlight the function y2. Press F4 to remove the √ to the left of y2.

107

- Highlight the function $y3$. Press $\boxed{\text{F4}}$ to remove the ✓ to the left of $y3$.

- Now, press $\boxed{\text{F2}}$ $\boxed{6}$ to view the graph of $y = x$ in the standard viewing window.

  The line should make a 45° angle with the x-axis, but the apparent angle on this screen is less than 45°. (Note that the tick marks on the y-axis are much closer together than on the x-axis.)

Step 2: Change to square scaling.
Press $\boxed{\text{F2}}$ $\boxed{5}$ for ZoomSqr. Now, you should see a true 45° angle. (Note that the tick marks on the x- and y-axes are equally spaced.)

Step 3. Clear the stored functions from your calculator.
Press $\boxed{\diamond}$ $\boxed{\text{Y=}}$. Highlight a stored function and press $\boxed{\text{CLEAR}}$. Repeat for each stored function.

Note: *You can turn the functions for y1 and y2 back on by repeating Step 1. This time the process will restore the checks to the left of y1 and y2.*

That's it! You have completed the tutorial. Now practice and experiment on your own with the calculator until you feel comfortable with these basic operations. The remainder of this guide will introduce new techniques and occasionally review techniques from the tutorial as they are needed, chapter by chapter, for your work in *Precalculus: Concepts in Context, 2e.*

Chapter-by Chapter Guide

CHAPTER 1

Returning to and Clearing the Home Screen.

If you want to exit any screen, press $\boxed{\text{2nd}}$ $\boxed{\text{QUIT}}$ (same key as $\boxed{\text{ESC}}$). On the TI-89/92Plus, you will return to the home screen. On the TI-Voyage 200, you will return to the main menu; press $\boxed{\text{2nd}}$ $\boxed{\text{CALC HOME}}$ to return to the home screen. To clear the home screen, press $\boxed{\text{F1}}$ $\boxed{\text{8}}$ for Clear Home. Press $\boxed{\text{CLEAR}}$ to clear the command line.

Plotting Points

You can use your calculator to plot the Fahrenheit-Celsius data from Lab 1A. Then graph your guess for the formula that relates degrees Fahrenheit to degrees Celsius. This will allow you to check how closely the function specified by your formula follows the pattern of the data.

Example: Plot the data in Table 1 and then overlay the graph of $y = 18x + 85$.
   You'll work on this problem in four steps. In Step 1, you'll enter the data; in Step 2, you'll plot these data; in Step 3, you'll add a line to your plot; and in Step 4, you'll turn off the data plot.

Sample Data	
x	y
-2	40
-1	60
1	100
3	140

Table 1. Sample Data

Step 1: Enter the data. Here's how.
   • On the TI-89/92Plus, press $\boxed{\text{APPS}}$ $\boxed{\text{6}}$ for the Data/Matrix Editor. On the TI-Voyage 200, highlight the Data/Matrix icon on the main menu (if necessary, press $\boxed{\text{2nd}}$ $\boxed{\text{QUIT}}$ to get to the main menu.) and then press $\boxed{\text{ENTER}}$.

   • Press $\boxed{\text{3}}$ for New. Your screen should match Figure 9.

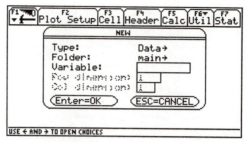

Figure 9. The Data/Matrix Screen

- Press the down arrow key to move the cursor to the box opposite Variable. Name your data set *sample*. (On a TI-89, you do not need to press $\boxed{\text{alpha}}$ before pressing the keys for the letters. Your calculator automatically locks the alpha-key because it expects you to enter a word. If you want to include a number in your variable name, press $\boxed{\text{alpha}}$ to release the alpha lock.) Press $\boxed{\text{ENTER}}$ twice to access the data table.

- Press the up arrow key to highlight the cell above c1. (See Figure 10.) Enter the name of the first variable: press $\boxed{\text{X}}$ $\boxed{\text{ENTER}}$.

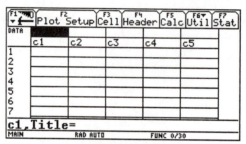

Figure 10. Data Table

- Press the down arrow key to highlight the cell directly beneath c1. Then enter the first number, − 2, and press $\boxed{\text{ENTER}}$. Enter the remaining 3 numbers from the x-column. Press $\boxed{\text{ENTER}}$ after keying in each number.

- Next use your arrow keys to highlight the cell above c2. Enter the name of the second variable: press $\boxed{\text{Y}}$ $\boxed{\text{ENTER}}$.

- Press the down arrow key to highlight the cell directly below c2. Enter the y-data. Your screen should match Figure 11. (To see the first row of data, you may need to use the up arrow key to scroll up.)

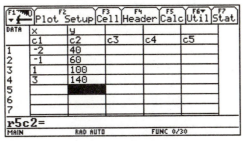

Figure 11. Entering Data

Step 2: Make a scatterplot of the data.
- Press $\boxed{F2}$ for Plot Setup.

- Plot 1 should be highlighted.
    Press $\boxed{F1}$ to define Plot 1.
    The Plot Type should be set to scatter.
    Use the down arrow key to highlight the entry opposite Mark. Press the left arrow key to bring down the choices. Press $\boxed{4}$ for Square.
    In the box opposite x, enter c1. (On a TI-89, you will need to press $\boxed{C}$ $\boxed{alpha}$ $\boxed{1}$ for c1).
    In the box opposite y, enter c2 and press $\boxed{ENTER}$. (See Figure 12.)
    Press $\boxed{ENTER}$ twice more, once to define the plot and a second time to return to the Data/Matrix table.

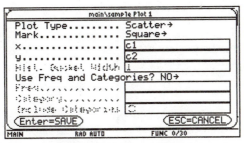

Figure 12. Setting up Plot 1

- Press $\boxed{\diamond}$ $\boxed{Window}$ $\boxed{F2}$ for Zoom. Press the down arrow until ZoomData is highlighted, and then press $\boxed{ENTER}$. Notice that the calculator automatically selected window settings that displayed all of the data. Your screen should look something like Figure 13.

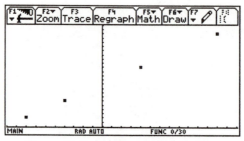

Figure 13. Scatterplot of Sample Data

Step 3: Graph the line and the scatterplot in the same viewing window.

• Press [◇] [Y=].

• Enter $18x + 85$ opposite $y1$. Press [ENTER].

• Press [◇] [GRAPH].

Step 4: Turn off the data plot.

Press [◇] [Y=] [F5] [5] for Data Plots Off. This will turn off Plot 1.

Warning! *If you fail to turn off Plot 1 off, your calculator will attempt to plot the data in c1 and c2 every time you graph. If you delete or change the data, your calculator will still try to graph Plot 1 the next time you press* [◇] [GRAPH]. *In that case, you are likely to get one of the error messages shown in Figure 14. So, remember to turn data plots off!*

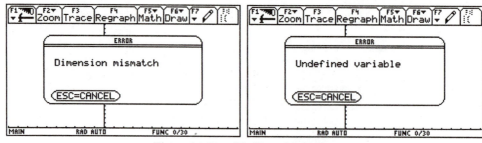

Figure 14. Possible Error Messages

Note: *Data variables take up memory. To erase the Data variable you created in the previous example, press* [2nd] [VAR-LINK] *(same key as* [—] *). Use the down arrow key to highlight* **sample**. *Then press* [←] [ENTER] *to delete the data variable.*

Making a Table of Values from a Formula

Your graphing calculator's table feature provides numeric information about a function. In the next example, you'll start with a column of values for the independent variable, x, and use the table feature to calculate the corresponding values for the dependent variable, y.

Example: Use your TI-89/92Plus/Voyage 200 to complete Table 2 for the function $y = 2x - 30$.

x	y
2.0	
6.0	
10.0	
14.0	
18.0	
22.0	
26.0	

Table 2. A Table of Function Values

Plan of action: First, you'll enter the function into the calculator. Then you'll set up the values for the independent variable, x. Notice that the minimum x-value in the table is 2.0 and that consecutive x-values are separated by increments of 4.0. Using the TABLE command, you'll be able to generate the values for the x- and y-columns.

- Press ◇ Y= . Clear any previously stored functions. Then enter the function $y = 2x - 30$ as $y1$ and press ENTER .

- Press ◇ TblSet (on the TI-89, same key as F4 ; on the TI92Plus/Voyage 200, same key as T) to access the TABLE SETUP screen.

- After entering 2 in the box opposite tblStart and 4 in the box opposite Δtbl, your screen should match the one in Figure 15. Press ENTER twice to save the settings and return to the previous screen.

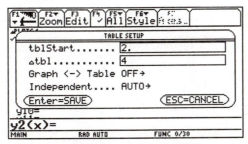

Figure 15. Setting Up a Table

- To view the table, press ◇ TABLE (on the TI-89, same key as F5 ; on the TI-92Plus/Voyage 200, same key as Y).

Using the up and down arrows, you can scroll up or down in the table. For example, use the down arrow key to find the value of $y1$ that corresponds to an x-value of 62. (You should get

94.) Next, use the up arrow key to find two x-values in the table for which the dependent variable changes signs. (You should get x-values of 14 and 18.)

Computing the Value of a Function

Example: Find the value of the function $f(x) = 3x^2 + 6x - 7$ for $x = 5$ and $x = 8$. In other words, find the values of $f(5)$ and $f(8)$.

- Press ◇ Y= . Enter this function as $y1$ and then return to the home screen. (Return to the home screen on the TI-89/92Plus by pressing 2nd QUIT , and on the TI-Voyage 200 by pressing 2nd CALC HOME .)

- Enter $y1(5)$ on the entry line as follows: press Y 1 (5) and then press ENTER . You should get 98.

- Compute $y1(8)$ as follows. Press the right arrow key to remove the highlighting on the entry line. The cursor will be positioned to the right of $y1(5)$. Press the left arrow key once to move the cursor next to 5. Press ← and then 8 ENTER . You should get 233.

Example: Find the average rate of change of $f(x) = 3x^2 + 6x - 7$ from $x = 5$ to $x = 7$.

From the previous example, you should already have the formula for $f(x)$ stored as $y1$. Now, enter the formula for the average rate of change on the entry line and press ENTER . Your entry line should match the one in Figure 16.

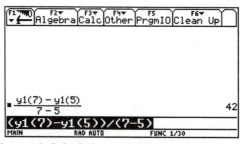

Figure 16. Calculating an Average Rate of Change

Adjusting the Window Settings for Square Scaling

The viewing screen on your calculator is a rectangle. Therefore, if you use the standard window, the tick marks on the y-axis will be closer together than those on the x-axis. For square scaling, you want the distance between 0 and 1 on the x-axis to be the same as the distance between 0 and 1 on the y-axis.

Example: The graph of $f(x) = \sqrt{25 - x^2}$ forms the top half of a circle. Graph $f(x) = \sqrt{25 - x^2}$ in the standard window and then switch to square scaling.

You'll do this problem in two steps. In Step 1, you'll enter the function and graph it in the standard viewing window. In Step 2, you'll change to square scaling.

Step 1: Enter $\sqrt{25 - x^2}$ as $y1$ and then graph it in the standard viewing window.
- Press $\boxed{\diamond}$ $\boxed{\text{Y=}}$. Clear any previously stored functions. Then enter $\sqrt{25 - x^2}$ as $y1$. (Remember there should be both left and right parentheses.)

- Press $\boxed{\text{F2}}$ $\boxed{6}$ for ZoomStd.
  Observe the spacing between the tick marks on the x- and y-axes. Notice that the tick marks on the y-axis are closer together than the tick marks on the x-axis. This graph does not appear to be a semicircle.

- Press $\boxed{\diamond}$ $\boxed{\text{WINDOW}}$ to view the window settings of the standard window.

Step 2: Change to a square viewing window.
- Press $\boxed{\text{F2}}$ $\boxed{5}$ to select ZoomSqr. When the graph appears on your screen, observe the equal distance between tick marks on the two axes. In this window, the graph is shaped like a semicircle.

- Press $\boxed{\diamond}$ $\boxed{\text{WINDOW}}$ and note the changes in the window settings.

Example: Add the graph of $g(x) = -\sqrt{25 - x^2}$ to the previous graph.
  This function is the opposite of the previous function. So, you can define it in terms of $y1$. Here's how.

- Press $\boxed{\diamond}$ $\boxed{\text{Y=}}$. Your screen should match Figure 17. The cursor, in this case a black square, should be opposite $y2$.

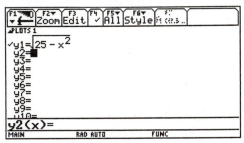

Figure 17. Contents of Y= Editor

- Press $\boxed{\text{(-)}}$ $\boxed{\text{Y}}$ $\boxed{1}$ $\boxed{\text{(}}$ $\boxed{\text{X}}$ $\boxed{\text{)}}$ $\boxed{\text{ENTER}}$.

- To see the graph of both halves of a circle, press $\boxed{\diamond}$ $\boxed{\text{GRAPH}}$.

CHAPTER 2

Finding Values of a Recursive Function

Example: Suppose that $P(0) = 2$ and that $P(t+1) = P(t) + 5$. Use your calculator to find $P(1)$, $P(2)$, $P(3)$, and $P(4)$.

There are several ways to solve this problem. Here's one that uses your calculator's answer feature.

- From the home screen, enter the value for $P(0)$: press $\boxed{2}$ $\boxed{\text{ENTER}}$.

- Since $P(1) = P(0) + 5$, press $\boxed{+}$ $\boxed{5}$ $\boxed{\text{ENTER}}$ to add 5 to the value of $P(0)$.

- Press $\boxed{\text{ENTER}}$ three more times to find the values of $P(2)$, $P(3)$, and $P(4)$. (Check that your answers match those in Figure 18.)

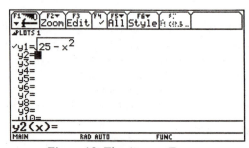

Figure 18. The Answer Feature

Fitting a Line to Data

If your data, when plotted, lie exactly on a line, you can use algebra to determine the equation of the line. However, real data seldom fall precisely on a line. Instead, the plotted data may exhibit a roughly linear pattern. The least squares line (also called the regression line) is a line that statisticians frequently use when describing a linear trend in data.

Example: Use the least squares line to describe the linear pattern in the data displayed in Table 3.

x	y
-3.0	-6.3
-2.0	-2.8
1.2	2.0
2.0	4.1
3.1	5.0
4.2	7.2

Table 3. Data With Linear Trend

You'll tackle this problem in three steps. In Step 1, you'll enter the data. In Step 2, you'll determine the equation of the least squares line. Then in Step 3, you'll plot the data and graph the line.

116

Step 1: Erase any stored functions and then enter the data.

- Press ◇ Y= and erase any stored functions.

- Press APPS . On the TI-89/92Plus, press 6 ; on a TI-Voyage 200, highlight the Data/Matrix icon and press ENTER . Then press 3 for new.

- Enter a name for the Data variable in the empty box. Then press ENTER twice to bring up a blank data table.

- Enter your data, placing the data from the x-column in c1 and the data from the y-column in c2. (For more specific details, refer to *Plotting Points* in Chapter 1, starting on page 109.)

Step 2: Next, we'll find the equation for the least squares line. Your calculator will determine the values for the slope and intercept of $y = ax + b$.

- Press F5 to select the Calc menu.

- Next, press the right arrow key to show the choices for Calculation Type. Press 5 to select LinReg. (*If you want to fit a different function, such as an exponential function or a quadratic function, see the note at end of this section.*)

- Complete the remainder of the screen as follows:
  Enter c1 for x and c2 for y. (On the TI-89, you will have to use the alpha key, both to enter letters and to release the alpha lock.)

  You'll want to store the regression equation: press the down arrow key to highlight the entry to the right of Store RegEQ to. Press the left arrow key. Select $y1(x)$ for the storage location of the regression equation by highlighting it and pressing ENTER . When you are finished, your screen should match Figure 19.

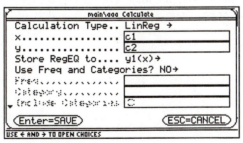

Figure 19. Setting Up a Regression

- Press ENTER . Your calculator will display the values for the slope and y-intercept of the regression (least-squares) line. Press ENTER again to return to the data table.

117

Step 3: Make a scatterplot of the data and overlay a graph of the least squares line.
- Press $\boxed{F2}$. Plot 1 should be highlighted. Press $\boxed{F1}$ and define Plot 1. (Refer to *Plotting Points* in Chapter 1, starting on page 109.)

- After you have defined Plot 1, press $\boxed{\diamond}$ $\boxed{Y=}$. Plot 1 should be checked and the regression equation should be stored as $y1$. Now press $\boxed{F2}$ $\boxed{9}$ for ZoomData. (On a TI-89, you will have to scroll down with the down arrow key to see it.) See Figure 20.

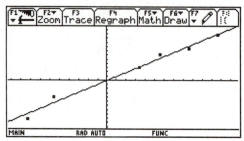

Figure 20. The Least-Squares Line

Warning! *Remember to turn off Plot 1 before beginning a new example. Press* $\boxed{\diamond}$ $\boxed{Y=}$ $\boxed{F5}$ $\boxed{5}$ *for Data Plots Off.*

Note: *You can adapt the instructions from Step 2 above to fit other functions to data. For example, if you want to fit an exponential function (instead of a linear function), select ExpReg (instead of LinReg) for the Calculation Type. QuadReg will fit a quadratic function, CubicReg will fit a cubic function, SinReg will fit a sinusoidal function, and so forth.*

Graphing an Exponential Function

Example: Graph $p(x) = 200(1.09)^x$ over the interval from $x = 0$ to $x = 10$. Use ZoomFit to choose the window settings for $ymin$ and $ymax$. Then change the window settings and graph the function over the interval from $x = 0$ to $x = 50$
- Press $\boxed{\diamond}$ $\boxed{Y=}$ and clear any stored functions.

- Enter the formula for $p(x)$ as $y1$. (Use $\boxed{\wedge}$ for the exponent.)

- Set the x-interval. Press $\boxed{\diamond}$ $\boxed{WINDOW}$. Set $xmin = 0$, $xmax = 10$, $xscl = 0$ and $yscl = 0$ (to turn off the tick marks on the axes).

- Next, let your calculator set the values for $ymin$ and $ymax$. Here's how. Press $\boxed{F2}$. Then press the down arrow key until ZoomFit is highlighted and press $\boxed{ENTER}$. Your graph should resemble the one in Figure 21.

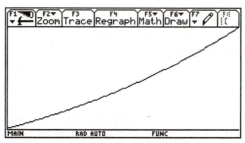

Figure 21. A ZoomFit Window

- Check the window settings: press ⬦ | WINDOW |.

- Change the window settings for *x*max to 50, respectively. Then press | F2 |, highlight ZoomFit and press | ENTER |. In this window, the graph should appear to curve upward more steeply than the previous graph.

Finding the Coordinates of Points of Intersection

Example: Find the points where the graphs of $f(x) = -4x + 15$ and $g(x) = -2x^2 + 2x + 12$ intersect. You'll tackle this problem in three steps. In Step 1, you'll graph the functions. You'll find the coordinates of one of the points of intersection in Step 2, and the other in Step 3.

Step 1: Graph the functions in a viewing window that gives a clear view of the points of intersection.
- Enter the formulas for $f(x)$ and $g(x)$ into your calculator as $y1$ and $y2$, respectively. Erase any other previously stored functions.

- Adjust the window settings so that you can see both points of intersection. (*Hint:* you might start with the standard viewing window, | F2 | | 6 |, and then adjust the window settings after looking at the graph.)

Step 2: Approximate the coordinates of one of the points of intersection.
- Press | F5 | to access the Math menu and then press | 5 | for intersection. The image on your screen should be similar to Figure 22.

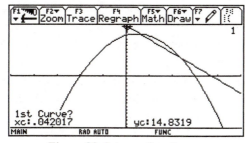

Figure 22. Intersecting Curves

- Your calculator wants to know which curve you want to call the 1st Curve. Try pressing the up and down arrow keys. The cursor will jump back and forth from the line to the parabola. Position the cursor on the line. (We'll designate the line as the first curve since its equation is entered as $y1$.) Then press $\boxed{\text{ENTER}}$. The cursor should then jump to the parabola. Press $\boxed{\text{ENTER}}$ to designate the parabola as the second curve.

- Approximate the point of intersection having the smaller x-coordinate as follows. You'll have to specify a narrow interval about this point of intersection. To specify the lower bound of this interval, use the left or right arrow keys to position the cursor slightly to the left of this point of intersection and then press $\boxed{\text{ENTER}}$. To specify the upper bound of this interval, use the right arrow key to move the cursor slightly to the right of this point of intersection (but not so far right that the second intersection point also lies in the interval). Then press $\boxed{\text{ENTER}}$. The coordinates of the first intersection point appear at the bottom of your screen.

Step 3: Find the coordinates of the second point of intersection.
Repeat the process outlined in Step 2 to find the coordinates of the second intersection point.

If you have done everything correctly, you will find that the two graphs intersect at approximately (0.634, 12.464) and (2.366, 5.536).

Graphing Piecewise-Defined Functions

Example: Graph the piecewise-defined function $f(x) = \begin{cases} x - 4, & \text{if } x > 4 \\ -x + 4, & \text{if } x \leq 4 \end{cases}$.

The graph of $f(x)$ consists of two half-lines pieced together. You'll want the graph of $y = x - 4$ when x-values are greater than 4 and $y = -x + 4$ when x-values are less than or equal to 4. In order to accomplish this, you'll use the **when** command. Here's the syntax of the command:

when(condition, expression 1, expression 2).

Here's how this command works. Expression 1 is calculated if the condition is true and expression 2 is calculated if the condition is false. Next, you'll use the **when** command to graph $f(x)$ in the standard viewing window.

- Press $\boxed{\diamond}$ $\boxed{\text{Y=}}$ and clear any stored functions. If necessary, move the cursor opposite $y1$.

- You can access the **when** command through the CATALOG. On the TI-89, press $\boxed{\text{CATALOG}}$; on the TI-92Plus/Voyage 200, press $\boxed{\text{2nd}}$ $\boxed{\text{CATALOG}}$ (same key as $\boxed{2}$). Scroll down to the w's by pressing $\boxed{\text{W}}$ (on the TI-89, same key as $\boxed{\cdot}$, no need to press $\boxed{\text{alpha}}$ first). The triangle marker should be opposite **when(**. Press $\boxed{\text{ENTER}}$.

- Press $\boxed{\text{X}}$ $\boxed{\text{2nd}}$ $\boxed{>}$ (same key as $\boxed{\cdot}$) $\boxed{4}$ $\boxed{,}$ (the key above $\boxed{9}$) to enter the condition.

- Press $\boxed{\text{X}}$ $\boxed{-}$ $\boxed{4}$ $\boxed{,}$ to enter expression 1.

- Press $\boxed{(-)}$ $\boxed{X}$ $\boxed{+}$ $\boxed{4}$ $\boxed{)}$ to complete the command.

- Press $\boxed{\text{ENTER}}$. Your screen should match Figure 23.

- Press $\boxed{\text{F2}}$ $\boxed{6}$. Check that your graph matches the one in Figure 24.

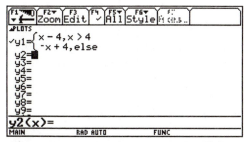 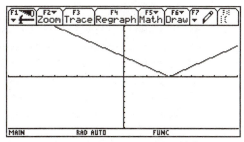

Figure 23. Entering the Formula for $f(x)$ Figure 24. The Graph of $f(x)$

Example: Graph $g(x) = \begin{cases} -2x, & \text{if } x < 1 \\ -2 + 3(x-1), & \text{if } 1 \le x < 3 \\ 4, & \text{if } x \ge 3 \end{cases}$.

This is a bit tricky. You'll have to nest two when statements. Here's how:

$$\text{when}(x < 1, \ -2x, \ \text{when}(x < 3, \ -2 + 3(x-1), 4))$$

So, for $x < 1$, $y = -2x$ is graphed. But for $x \ge 1$ (these are the x-values that make the condition $x < 1$ false), when$(x < 3, \ -2 + 3(x-1),4)$ is graphed.

Remember, if you get stuck in some menu along the way, you can always press $\boxed{\text{2nd}}$ $\boxed{\text{QUIT}}$ and then start again.
- Press $\boxed{\diamond}$ $\boxed{Y=}$ and erase any stored functions.

- Enter *when(x < 1, -2x, when(x < 3, -2 + 3(x-1),4))* as $y1$. (Refer to the previous example if you need help.)

- Press $\boxed{\diamond}$ $\boxed{\text{GRAPH}}$ or $\boxed{\text{F2}}$ $\boxed{6}$ to graph $g(x)$. See Figure 25.

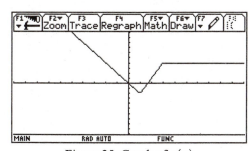

Figure 25. Graph of $g(x)$

Example: Graph the greatest integer function, $[[x]]$ in the standard viewing window. (See Chapter 2, Exercise 46 in *Precalculus: Concepts in Context, 2e* for a definition of this function.)

The greatest integer function is one of your calculator's built-in functions. Its calculator name is int(x). You can either type in the name or access it from the catalog.

- Press ◇ Y= and clear any stored functions. Position the cursor opposite $y1$.

- To enter the function, press CATALOG on the TI-89 and 2nd CATALOG on the TI-92Plus/Voyage 200. Press I (on the TI-89, same key as 9 , no need to press alpha first). Then use the down arrow key to move the triangle marker opposite **int(**. Press ENTER X) ENTER .

- Press F2 6 to view the graph. Your graph should resemble Figure 26.

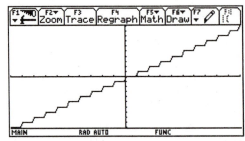

Figure 26. Graph of the Greatest Integer Function

The vertical line segments that appear on your screen cannot be part of the graph of a function. (Why?) For example, the nearly vertical line segment at $x = 3$ is due to the fact that your graphing calculator plotted a point on the left of $x = 3$, and a point on the right of $x = 3$, and then connected the points with a straight line segment. One solution to this problem is to change your graphing style from Line to Dot. That's what you'll do next.

- Press ◇ Y= . Press the up arrow key to highlight int(x). Then, on the TI-89, press 2nd F6 (same key as F1); on the TI-92Plus/Voyage 200 press F6 . This will bring down the Style menu. Press 2 for Dot or highlight 2 and press ENTER . Then press ◇ Graph. . The nearly vertical line segments should now be gone.

- Return the style for $y1$ to Line.

CHAPTER 3

In Chapter 3 you will be investigating the effect that certain algebraic modifications, such as adding a constant to the input variable, have on the graph of a function. You'll want to experiment using several different functions. We've provided some functions and algebraic modifications that you might want to consider.

Using Parentheses

Warning! *When you want to apply a function to an expression, you must enclose the entire expression in parentheses. For built-in functions (such as the square root or the sine function or the greatest integer function), the TI-89/92Plus/Voyage 200 will automatically insert the left parenthesis and you will need to add the right parenthesis.*

Example: Graph $y = \sqrt{x+2}$ in the window $[-5,5] \times [-5,5]$.
- Press $\boxed{\diamond}$ $\boxed{\text{Y=}}$. Then enter $y = \sqrt{x+2}$ as $y1$ by pressing $\boxed{\text{2nd}}$ $\boxed{\sqrt{}}$ $\boxed{\text{X}}$ $\boxed{+}$ $\boxed{2}$ $\boxed{)}$ $\boxed{\text{ENTER}}$.

- Press $\boxed{\diamond}$ $\boxed{\text{WINDOW}}$ and adjust the settings for a $[-5,5] \times [-5,5]$ window.

- Press $\boxed{\diamond}$ $\boxed{\text{GRAPH}}$. Your graph should resemble the one in Figure 27.

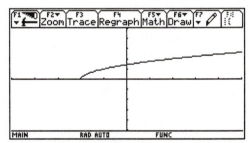

Figure 27. A Member of the Square Root Family

- Return to the Y= editor. Re-enter the function but this time omit the right parenthesis. Here's how. Press $\boxed{\diamond}$ $\boxed{\text{Y=}}$. Highlight the function and then press $\boxed{\text{ENTER}}$. Press the right arrow key and then press the backspace key $\boxed{\leftarrow}$ to erase the right parenthesis. Press $\boxed{\text{ENTER}}$. You should get an error message that tells you "Missing)". Press $\boxed{\text{ENTER}}$ and you will be taken to the location where the parenthesis is expected. Just press $\boxed{\text{CLEAR}}$ and then $\boxed{\text{ENTER}}$ to erase this function.

Graphing Functions Involving the Absolute Value

Example: Graph $f(x) = |x|$. Then add the graph of $g(x) = \frac{|x|}{x}$. Use a $[-5,5] \times [-5,5]$ window.
You can access the absolute value function from the catalog or by keying in the name **abs**. In this example, you'll do both.

123

- Press $\boxed{\diamond}$ $\boxed{Y=}$ and clear any stored functions.

- Enter $y = |x|$ as $y1$. Here's how:
    On the TI-89 press $\boxed{2\text{nd}}$ $\boxed{\text{alpha}}$ (to lock the alpha key) $\boxed{A}$ $\boxed{B}$ $\boxed{S}$ $\boxed{\text{alpha}}$ (to unlock the alpha key) $\boxed{(}$ $\boxed{X}$ $\boxed{)}$ $\boxed{\text{ENTER}}$.
    On the TI-92Plus/Voyage 200, press $\boxed{A}$ $\boxed{B}$ $\boxed{S}$ $\boxed{(}$ $\boxed{X}$ $\boxed{)}$ $\boxed{\text{ENTER}}$.

- If you worked through the previous example, you already have the correct window settings. (Otherwise press $\boxed{\diamond}$ $\boxed{\text{WINDOW}}$ and adjust the window settings.) Press $\boxed{\diamond}$ $\boxed{\text{GRAPH}}$ to view the V-shaped graph of the absolute value function.

- Next, enter the formula for $g(x)$ as $y2$.
    Press $\boxed{\diamond}$ $\boxed{Y=}$. The cursor should be opposite $y2$.
    On the TI-92Plus/Voyage 200, press $\boxed{2\text{nd}}$ $\boxed{\text{CATALOG}}$; on a TI-89, press $\boxed{\text{CATALOG}}$.
    Press $\boxed{A}$ and, if necessary, move the triangle to point toward **abs(**. Press $\boxed{\text{ENTER}}$.
    Press $\boxed{(}$ $\boxed{X}$ $\boxed{)}$ $\boxed{\div}$ $\boxed{X}$ $\boxed{\text{ENTER}}$.

- Press $\boxed{\diamond}$ $\boxed{\text{GRAPH}}$ to view the two graphs.

Note: *The function $g(x) = \frac{|x|}{x}$ tells you the sign of the input. The value of $g(x)$ is -1 for negative inputs and $+1$ for positive inputs. It is undefined at zero.*

Example: Use your calculator to evaluate $f(x) = |x|$ and $g(x) = \frac{|x|}{x}$ at $x = 0$.
From the previous example, you should already have $f(x)$ and $g(x)$ stored as $y1$ and $y2$, respectively. Begin these calculations from the home screen.

- First, evaluate $f(0)$: press $\boxed{Y}$ $\boxed{1}$ $\boxed{(}$ $\boxed{0}$ $\boxed{)}$ $\boxed{\text{ENTER}}$. You should get 0.

- Next, evaluate $g(0)$:
    Press the right arrow key.
    Press the left arrow key to move the cursor to the right of 1.
    Press $\boxed{\leftarrow}$ $\boxed{2}$ $\boxed{\text{ENTER}}$. Do you know why $g(0)$ is undefined?

Note: *If you modify $g(x)$ by setting $g(0) = 0$, then you get the sign function. You can access the sign function from your calculator's catalog. This function is defined by the following piecewise formula:*

$$\text{sign}(x) = \begin{cases} \frac{|x|}{x}, & \text{if } x \neq 0 \\ 0, & \text{if } x = 0 \end{cases}.$$

The sign function is not to be confused with the sine function, the next topic.

Graphing the Sine Function

Example: Graph $y = \sin(x)$ in the trig viewing window.
  You will learn more about this function in Chapter 6 of *Precalculus: Concepts in Context, 2e.*
  - First check that your calculator is in Radian Mode: Press $\boxed{\text{MODE}}$. Check the setting for Angle. It should read RADIAN. If it does not, use the down arrow key to highlight the setting for Angle, then press the right arrow key. Use the up arrow key to highlight Radian and press $\boxed{\text{ENTER}}$ twice.

  - Press $\boxed{\diamond}$ $\boxed{\text{Y=}}$. Enter $y = \sin(x)$ for $y1$ as follows:
    On the TI-89, press $\boxed{\text{2nd}}$ $\boxed{\text{SIN}}$ $\boxed{\text{X}}$ $\boxed{)}$ $\boxed{\text{ENTER}}$.
    On the TI-92/Voyage 200, press $\boxed{\text{SIN}}$ $\boxed{\text{X}}$ $\boxed{)}$ $\boxed{\text{ENTER}}$.

  - Press $\boxed{\text{F2}}$ $\boxed{7}$ to view the graph of $y = \sin(x)$ in the trig viewing window.

Graphing a Family of Functions

Using your calculator's list capabilities, you can substitute each value in a given list for a constant in an algebraic formula. This feature allows you to graph an entire family of functions quickly. On the TI-89/92Plus/Voyage 200 you specify a list by enclosing the members of the list in brackets: { }.

Example: Graph the family of quadratic functions $y = (x + 1)^2$, $y = (x + 2)^2$, and $y = (x + 3)^2$ in the window $[-5, 5] \times [-1, 10]$.
  - Press $\boxed{\diamond}$ $\boxed{\text{WINDOW}}$ and adjust the settings for a $[-5, 5] \times [-1, 10]$ window.

  - Press $\boxed{\diamond}$ $\boxed{\text{Y=}}$ and clear any stored functions. Position the cursor opposite $y1$.

  - Enter the three functions by specifying the constants 1, 2, and 3 in a list as follows:
    Press $\boxed{(}$ $\boxed{\text{X}}$ $\boxed{+}$.
    Press $\boxed{\text{2nd}}$ $\boxed{\{}$ $\boxed{1}$ $\boxed{,}$ $\boxed{2}$ $\boxed{,}$ $\boxed{3}$ $\boxed{\text{2nd}}$ $\boxed{\}}$.
    Press $\boxed{)}$ $\boxed{\wedge}$ $\boxed{2}$ $\boxed{\text{ENTER}}$.

  - Press $\boxed{\diamond}$ $\boxed{\text{GRAPH}}$ and watch as the three functions are graphed one after the other.

Expanding Expressions and Solving Equations

Your TI-89/92Plus/Voyage 200 has symbolic manipulation capabilities. It can expand and factor expressions and solve inequalities and equations.

Example: Expand $3(x + 1)^2 - 2(x + 1) + 5$.
  - If necessary, press $\boxed{\text{2nd}}$ $\boxed{\text{QUIT}}$ (also press $\boxed{\diamond}$ $\boxed{\text{CALC HOME}}$ if you are using the TI-Voyage 200) to bring up the home screen. Clear your home screen (so you can start fresh for this example): press $\boxed{\text{F1}}$ $\boxed{8}$. Clear the entry line by pressing $\boxed{\text{CLEAR}}$.

  - Press $\boxed{\text{F2}}$ $\boxed{3}$ for expand.

- Enter the expression you wish to expand, $3(x+1)^2 - 2(x+1) + 5$, and then press $\boxed{)}$ $\boxed{\text{ENTER}}$. Your screen should match Figure 28. Note the expanded expression is $3x^2 + 4x + 6$.

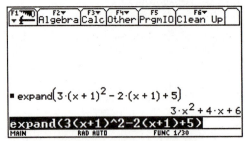

Figure 28. Expanding an Expression

Example: Solve $2x - 6 = 3x + 5$.
- Press $\boxed{\text{F2}}$ $\boxed{1}$ for solve.

- Enter the equation you want to solve: Press $\boxed{2}$ $\boxed{X}$ $\boxed{-}$ $\boxed{6}$ $\boxed{=}$ $\boxed{3}$ $\boxed{X}$ $\boxed{+}$ $\boxed{5}$ $\boxed{,}$ $\boxed{X}$ $\boxed{)}$ $\boxed{\text{ENTER}}$. You should get $x = -11$ for your solution.

Composing Functions

Example: Suppose $f(x) = 2x - 7$ and $g(x) = 5x^2$. Find $f(g(x))$.

On a TI-92, you may get an error message: Circular Definition. However, you will be able to use this process to evaluate compositions at specific x-values.

- Press $\boxed{\diamond}$ $\boxed{\text{Y=}}$. Erase any previously stored functions. Then enter the formula for $f(x)$ as $y1$ and the formula for $g(x)$ as $y2$. Then return to the home screen by pressing $\boxed{\text{2nd}}$ $\boxed{\text{QUIT}}$ on the TI-89/92Plus and $\boxed{\diamond}$ $\boxed{\text{CALC HOME}}$ on the TI-Voyage 200.

- Enter the composition on the command line: $\boxed{\text{Y}}$ $\boxed{1}$ $\boxed{(}$ $\boxed{\text{Y}}$ $\boxed{2}$ $\boxed{(}$ $\boxed{\text{X}}$ $\boxed{)}$ $\boxed{)}$ $\boxed{\text{ENTER}}$. The composed function will appear above the command line. (See Figure 29.)

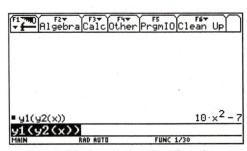

Figure 29. Composing Two Functions

Warning! *Always check that there are as many right parentheses as left parentheses. If your left and right parentheses don't match up, you'll get an error message.*

126

CHAPTER 4

Graphing Exponential Functions Involving e

There is one base for exponential functions that is so common that it has its own function key: $\boxed{e^x}$.

Example: Graph $f(x) = e^x$ in the window $[-3, 3] \times [-1, 12]$.
- Press $\boxed{\diamond}$ $\boxed{Y=}$ and clear any previously stored functions.

- Enter $f(x)$ as $y1$: .
  On the TI-89, press $\boxed{\diamond}$ $\boxed{e^x}$ (same key $\boxed{X}$). Then press $\boxed{X}$ $\boxed{)}$ $\boxed{ENTER}$.
  On the TI-92Plus/Voyage 200, press $\boxed{2nd}$ $\boxed{e^x}$ (same key as $\boxed{LN}$). Then press $\boxed{X}$ $\boxed{)}$ $\boxed{ENTER}$.

- Press $\boxed{\diamond}$ $\boxed{WINDOW}$ and adjust the settings for a $[-3, 3] \times [-1, 12]$ window.

- Press $\boxed{\diamond}$ $\boxed{GRAPH}$ to view the graph.

Example: Graph $g(x) = e^{-\frac{x}{2}}$ by editing the function from the previous example.
- Press $\boxed{\diamond}$ $\boxed{Y=}$.

- Press the up arrow key to highlight e^x and then $\boxed{ENTER}$ to return to the command line.

- Press the right arrow key to remove the highlighting and move the cursor to the end of the line.

- Press the left arrow key to move the cursor to the left side of x. Press $\boxed{(}$ $\boxed{(-)}$. Then press the right arrow key to move the cursor to the right side of x. Complete entering the function by pressing $\boxed{\div}$ $\boxed{2}$ $\boxed{)}$ $\boxed{ENTER}$.

- Press $\boxed{\diamond}$ $\boxed{GRAPH}$ to view the function in the same window as the previous example.

Do <u>not</u> erase $g(x)$ until you have worked through the next topic, *Turning Off the Axes*.

Turning Off the Axes

Sometimes it is helpful to view a graph without the presence of the x and y axes. In the last example, you may have noticed that the graph of $g(x)$ disappeared into the x-axis. Did the graph, in reality, disappear? In the next example, you'll turn off the axes and find out.

Example: View the graph of $g(x) = e^{-\frac{x}{2}}$ in the window $[-3, 3] \times [-1, 12]$.
  You should already have $g(x)$ stored as $y1$.

- Press $\boxed{\diamond}$ $\boxed{GRAPH}$.

- Press $\boxed{\text{F1}}$ $\boxed{9}$ for Format

- Use the down arrow key to highlight the setting for Axes. Press the right arrow key and then $\boxed{1}$ for Off and press $\boxed{\text{ENTER}}$.

- Turn the axes back on in preparation for the next example.

Graphing Logarithmic Functions

The natural logarithmic function (base e), $\ln(x)$, has its own function key $\boxed{\text{LN}}$. The common logarithmic function (base 10), $\log(x)$, can be accessed from the CATALOG or keyed in directly using the letter keys. Logarithmic functions of other bases can be graphed by dividing $\ln(x)$ or $\log(x)$ by the appropriate scaling factor.

Example: Graph $h(x) = \ln(x)$ and $g(x) = \log(x)$. Use the window $[-1, 12] \times [-2, 3]$.
- Press $\boxed{\diamond}$ $\boxed{\text{Y=}}$ and clear any previously stored functions.

- Enter $h(x)$ as $y1$:
  On the TI-89, press $\boxed{\text{2nd}}$ $\boxed{\text{LN}}$ $\boxed{\text{X}}$ $\boxed{)}$ $\boxed{\text{ENTER}}$.
  On the TI-92Plus/Voyage 200, press $\boxed{\text{LN}}$ $\boxed{\text{X}}$ $\boxed{)}$ $\boxed{\text{ENTER}}$.

- Enter $g(x)$ as $y2$:
  On the TI-89, press $\boxed{\text{2nd}}$ $\boxed{\text{ALPHA}}$ $\boxed{\text{L}}$ $\boxed{\text{O}}$ $\boxed{\text{G}}$ $\boxed{\text{ALPHA}}$ $\boxed{(}$ $\boxed{\text{X}}$ $\boxed{)}$ $\boxed{\text{ENTER}}$. (Note: pressing $\boxed{\text{2nd}}$ $\boxed{\text{ALPHA}}$ locks the alpha-key. Pressing $\boxed{\text{ALPHA}}$ unlocks the alpha-key.)
  On the TI-92Plus/Voyage 200, press $\boxed{\text{L}}$ $\boxed{\text{O}}$ $\boxed{\text{G}}$ $\boxed{(}$ $\boxed{\text{X}}$ $\boxed{)}$ $\boxed{\text{ENTER}}$.

- Press $\boxed{\diamond}$ $\boxed{\text{WINDOW}}$ and adjust the settings for a $[-1, 12] \times [-2, 3]$ window. Then press $\boxed{\diamond}$ $\boxed{\text{GRAPH}}$ to view the two graphs.

Approximating an Instantaneous Rate of Change

You can approximate the instantaneous rate of change of a function $f(x)$ at $x = a$ by evaluating the average rate of change of the function over a small interval containing a. For example, you might choose a symmetric interval about a of the form $(a - h, a + h)$ for some small value of h. Using this interval, your approximation of the instantaneous rate of change would be

$$\frac{f(a+h) - f(a-h)}{(a+h) - (a-h)} = \frac{f(a+h) - f(a-h)}{2h}.$$

This is called the central difference quotient, which can be computed using the command **nDeriv**. The syntax of the command is shown below:

$$\text{nDeriv(function, variable, }h)|\text{variable} = \text{value}$$

Example: Use the central difference quotient to approximate the instantaneous rate of change of the function $f(x) = e^{2x}$ at $x = 0, 1,$ and 2. Use $h = 0.01$.

- Press $\boxed{\diamond}$ $\boxed{Y=}$ and clear any previously stored functions.

- Enter e^{2x} for $y1$. (Be sure to enclose the exponent in parentheses.) Then return to the home screen (on the TI-89/92Plus, press $\boxed{2\text{nd}}$ $\boxed{\text{QUIT}}$; on the TI-Voyage 200, press $\boxed{\diamond}$ $\boxed{\text{CALC HOME}}$.

- Press $\boxed{F3}$ to access the Calc menu. Press the down arrow key to highlight **nDeriv(**. Then press $\boxed{\text{ENTER}}$.

- Press $\boxed{Y}$ $\boxed{1}$ $\boxed{(}$ $\boxed{X}$ $\boxed{)}$ $\boxed{,}$ $\boxed{X}$ $\boxed{,}$ $\boxed{.}$ $\boxed{0}$ $\boxed{1}$ $\boxed{)}$. Keep going.

- Complete the command as follows:

  On the TI-89, press $\boxed{|}$ $\boxed{X}$ $\boxed{=}$ $\boxed{0}$ $\boxed{\text{ENTER}}$.

  On the TI-92Plus/Voyage 200, press $\boxed{2\text{nd}}$ $\boxed{|}$ (same key as $\boxed{K}$) $\boxed{X}$ $\boxed{=}$ $\boxed{0}$ $\boxed{\text{ENTER}}$.

Your screen should be similar to Figure 30.

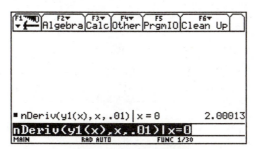

Figure 30. Using nDeriv

- Press the right arrow key to move the cursor to the end of the command line. Press $\boxed{\leftarrow}$ $\boxed{1}$ to replace the value 0 with the value 1. Then press $\boxed{\text{ENTER}}$. Finally, replace the value of 1 with the value 2. You should get rates of approximately 14.8 and 109.2, respectively.

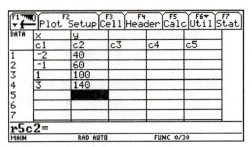

Figure 11. Entering Data

Step 2: Make a scatterplot of the data.
- Press F2 for Plot Setup.

- Plot 1 should be highlighted.
    Press F1 to define Plot 1.
    The Plot Type should be set to scatter.
    Use the down arrow key to highlight the entry opposite Mark. Press the left arrow key to bring down the choices. Press 4 for Square.
    In the box opposite x, enter c1. (On a TI-89, you will need to press C   alpha   1 for c1).
    In the box opposite y, enter c2 and press ENTER . (See Figure 12.)
    Press ENTER twice more, once to define the plot and a second time to return to the Data/Matrix table.

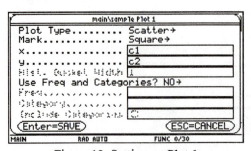

Figure 12. Setting up Plot 1

- Press ◇   Window   F2 for Zoom. Press the down arrow until ZoomData is highlighted, and then press ENTER . Notice that the calculator automatically selected window settings that displayed all of the data. Your screen should look something like Figure 13.

130

Note: *you also could have used your calculator's solve feature to find the solutions to* $6x^3 - 49x^2 + 50x + 168 = 0$*. However, then you may never have discovered the copy and paste features of your calculator.*

Example: Use the graph of $f(x)$ to find the x-intercepts. In other words, find the zeros of $f(x)$.

- Press $\boxed{\diamond}$ $\boxed{\text{Y=}}$. You may already have $f(x)$ stored as $y1$ from the previous example. If not, clear any stored functions and enter $f(x)$ as $y1$.

- Press $\boxed{\diamond}$ $\boxed{\text{WINDOW}}$ and set $x$min $= -3$ and $x$max $= 7$. Then press $\boxed{\text{F2}}$, use the down arrow key to highlight ZoomFit, and press $\boxed{\text{ENTER}}$. You should see a graph of $f(x)$ that shows three x-intercepts.

- To approximate the coordinates of the smallest x-intercept, press $\boxed{\text{F5}}$ $\boxed{2}$ for Zero. Press the left arrow key to move the cursor (a circle with cross-hairs) a little to the left of the negative x-intercept. Press $\boxed{\text{ENTER}}$. Then press the right arrow key to move the cursor a little to the right of this intercept and press $\boxed{\text{ENTER}}$. Read the approximate coordinates from the bottom of the screen.

- Repeat the process in the last bullet to find the approximate coordinates of the remaining two x-intercepts.

Check that the approximate values of the x-intercepts are close to the solutions to the previous example.

Finding Local Maxima or Minima

Example: Estimate the local maximum and local minimum of $f(x) = x^3 - 4x^2 + 2x - 4$.
You'll complete this example in three steps. In Step 1, you'll graph $f(x)$. In Step 2, you'll find the local minimum. In Step 3, you'll find the local maximum.

Step 1: Graph $f(x)$ using a viewing window that gives you a clear view of the two turning points (one peak and one valley) of the graph.

- Press $\boxed{\diamond}$ $\boxed{\text{Y=}}$ and clear any previously stored functions. Enter the formula for $f(x)$ as $y1$. Remember to press $\boxed{\text{ENTER}}$ after keying in the formula.

- Experiment with window settings until you find settings that show both turning points. The graph in Figure 32 shows one example.

131

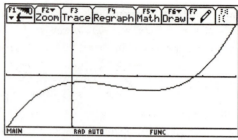

Figure 32. A Graph of $f(x)$

Step 2: Approximate the coordinates of the turning point associated with the local maximum (the y-coordinate of the peak on the graph) as follows:

• Press $\boxed{\text{F5}}$ for the Math menu and then $\boxed{4}$ for Maximum.

• You'll have to specify a narrow interval about the turning point associated with the local maximum (peak). Press the left-arrow key to move the cursor slightly to the left of the peak and then press $\boxed{\text{ENTER}}$.

• Next, press the right-arrow key to position the cursor slightly to the right of the turning point associated with the maximum and press $\boxed{\text{ENTER}}$.

• Read off the coordinates of this turning point at the bottom of your screen. The local maximum (y-coordinate of this turning point) should be approximately -3.73.

Step 3: Approximate the coordinates of the turning point associated with the local minimum (the y-coordinate of the valley on the graph).

• Press $\boxed{\text{F5}}$ $\boxed{3}$ for Minimum.

• Adapt the instructions for Step 2 to find the coordinates of the turning point associated with the local minimum (the valley). You should get a value for y that is close to -8.42.

Approximating Instantaneous Rates of Change

You can approximate the instantaneous rate of change of a function $f(x)$ at $x = a$ by evaluating the average rate of change of the function over a small interval containing a. For example, you might choose a symmetric interval about a of the form $(a - h, a + h)$ for some small value of h. Using this interval, your approximation of the instantaneous rate of change would be

$$\frac{f(a + h) - f(a - h)}{(a + h) - (a - h)} = \frac{f(a + h) - f(a - h)}{2h}.$$

This is called the central difference quotient.

Example: Use the central difference quotient to approximate the instantaneous rate of change of the function $f(x) = x^3 - 4x^2 + 2x - 4$ at $x = 1$ and $x = 3$. Use $h = .01$.

The formula for the central difference quotient using $a = 1$ and $h = .01$ is $\frac{f(1+.01)-f(1-.01)}{2(.01)}$. You'll do this problem in three steps. In Step 1, you'll enter the function. In Steps 2 and 3, you'll approximate the instantaneous rate of change of $f(x)$ at $x = 1$ and $x = 3$, respectively.

Step 1: Enter $f(x)$ as Y1 and return to the home screen.
- Press $\boxed{\diamond}$ $\boxed{\text{Y=}}$. If you completed the previous example, the formula for $f(x)$ may already be stored as $y1$. If not, enter the formula for $f(x)$ as $y1$.

- Return to the home screen.

Step 2: Approximate the instantaneous rate of change of $f(x)$ at $x = 1$.
Enter the formula for the central difference quotient using $x = 1$ and $h = .01$. Make sure you enter the entire formula before pressing enter. Here are the keystrokes.

- Enter the numerator:
  Press $\boxed{(}\boxed{\text{Y}}\boxed{1}\boxed{(}\boxed{1}\boxed{+}\boxed{.}\boxed{0}\boxed{1}\boxed{)}$.
  Press $\boxed{-}$.
  Press $\boxed{\text{Y}}\boxed{1}\boxed{(}\boxed{1}\boxed{-}\boxed{.}\boxed{0}\boxed{1}\boxed{)}\boxed{)}$.

- Press $\boxed{\div}$

- Enter the denominator: press $\boxed{(}\boxed{2}\boxed{\times}\boxed{.}\boxed{0}\boxed{1}\boxed{)}$

- Press $\boxed{\text{ENTER}}$. You should get approximately -3.0.

Step 3: Approximate the instantaneous rate of change of $f(x)$ at $x = 3$ by editing work from the previous step.
- Press the left or right arrow so you can edit the previous calculation.

- Change the two x-values of 1 in the numerator to 3: move the cursor to the right of 1 and then press $\boxed{\leftarrow}\boxed{3}$. After you have completed the edits, press $\boxed{\text{ENTER}}$. The result is approximately 5.0.

In this example, you approximated instantaneous rates of change directly from a formula. You might prefer using your calculator's **nDeriv** command instead. For more information, check out *Approximating Instantaneous Rates of Change* in Chapter 4, pages 128 and 129.

Graphing a Rational Function

Warning! *If the numerator or the denominator of a rational function consists of more than one term, you must enclose it in parentheses when you enter it into your calculator.*

Example: Graph $r(x) = \dfrac{x^2 - 1}{x - 3}$ in the standard viewing window. Then adjust window settings to display the key graphical features of this function.

• Press $\boxed{\diamond}$ $\boxed{Y=}$ and erase any previously stored functions.

• Enter $r(x)$ as $y1$: press $\boxed{(}$ $\boxed{X}$ $\boxed{\wedge}$ $\boxed{2}$ $\boxed{-}$ $\boxed{1}$ $\boxed{)}$ $\boxed{\div}$ $\boxed{(}$ $\boxed{X}$ $\boxed{-}$ $\boxed{3}$ $\boxed{)}$ and then $\boxed{\text{ENTER}}$. If your calculator is in Pretty Print Mode, the function for $y1$ should appear as written as shown above in the statement of the example.

• Press $\boxed{\text{F2}}$ $\boxed{6}$ to view the graph in the standard window. Your graph should look similar to the one in Figure 33.

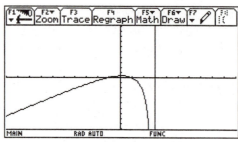

Figure 33. Graph of $r(x)$

The vertical line in the graph shown in Figure 33 indicates that this function has a vertical asymptote at $x = 3$. Remember, this line is not part of the graph of this function. Furthermore, because the domain of this function is all real numbers except $x = 3$, there is a branch of this function's graph that lies to the right of the line $x = 3$. In order to observe this branch, you will have to adjust the window setting for $y$max.

• Press $\boxed{\diamond}$ $\boxed{\text{WINDOW}}$. Change the setting for $y$max to 20. Then press $\boxed{\diamond}$ $\boxed{\text{GRAPH}}$. You should now see both branches of the graph along with the nearly vertical line that is not part of this function's graph.

• To remove the vertical line, you need to change the style from Line to Dot. Here's how. Press $\boxed{\diamond}$ $\boxed{Y=}$. Press the up arrow key to highlight $y1$'s formula. On the TI-89, press $\boxed{\text{2nd}}$ $\boxed{\text{F6}}$ (same key as F1) to bring down the Style menu. On the TI-92Plus/Voyage 200, press $\boxed{\text{F6}}$ to bring down the Style menu. Press $\boxed{2}$ (for Dot) or use the down arrow key to highlight Dot and press $\boxed{\text{ENTER}}$. Then press $\boxed{\diamond}$ $\boxed{\text{GRAPH.}}$. The vertical line should now be gone.

• Return the Style to Line for the next Example.

Do <u>not</u> clear this function from your calculator's memory until after you have completed the next topic, *Zooming Out*.

134

Zooming Out

Example: Graph the function $r(x) = \dfrac{x^2 - 1}{x - 3}$ in the standard viewing window and then zoom out by a factor of four several times.

For this example, observe what happens to the appearance of the graph of $r(x)$ as we "back away" by increasing the width and height of the viewing window. From the previous example, you should already have the formula for $r(x)$ stored as $y1$.

- Press $\boxed{\text{F2}}$ $\boxed{6}$ to graph the function in the standard viewing window.

- Press $\boxed{\diamond}$ $\boxed{\text{WINDOW}}$ to access the window settings. Set xScl = 0 and yScl = 0. (This turns off the tick marks that appear on the axes. If you skip this step, the axes will get crowded with tick marks when you zoom out.)

- Check the settings for the zoom factor. Press $\boxed{\text{F2}}$, highlight SetFactors, and press $\boxed{\text{ENTER}}$. Your screen should match Figure 34. (If it doesn't, change the settings for xFact and yFact to 4.) Press $\boxed{\text{ENTER}}$ to set the factors and return to the graph.

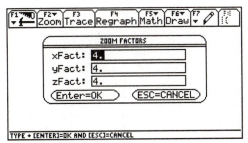

Figure 34. Settings For Zoom Factors

- Press $\boxed{\text{F2}}$ $\boxed{3}$ for ZoomOut. A blinking cursor marks the center of the zoom. (If you want to change the zoom center, use the arrow keys to move the cursor to a new location.) Press $\boxed{\text{ENTER}}$ to view the graph over wider x- and y-intervals.

- Press $\boxed{\text{F2}}$ $\boxed{3}$ again to zoom out a second time. Then press $\boxed{\text{ENTER}}$ to remove the cursor from your screen. (Compare your zoomed out graphs with those in Figure 35.)

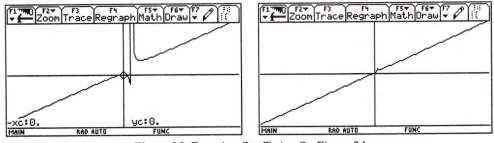

Figure 35. Zooming Out Twice On Figure 34

• Now, press $\boxed{\diamond}$ $\boxed{\text{WINDOW}}$ to observe the effect on the window settings of zooming out twice (by a factor of 4).

The default setting for ZoomOut widens both the x- and y-intervals by a factor of 4 each time that it is applied. In the previous example, you zoomed out twice. Therefore, the x- and y-intervals are 16 times wider than they were before you zoomed out.

In the next example, we'll change the default zoom settings in order to observe a graph of a function that begins to act like its horizontal asymptote.

Example: Graph $q(x) = \dfrac{5x^2 + 20x - 105}{2x^2 + 2x - 60}$ in the standard viewing window.

• Press $\boxed{\diamond}$ $\boxed{\text{Y=}}$. Clear any previously stored functions.

• Enter $q(x)$ as $y1$. (Remember to press $\boxed{\text{ENTER}}$ after keying in the function.) To view a graph of this function, press $\boxed{\text{F2}}$ $\boxed{6}$. If you have entered the function correctly, your graph should look like the one in Figure 36.

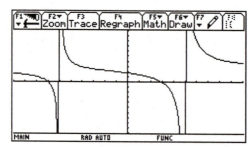

Figure 36. Graph of $q(x)$.

Based on this graph, you may suspect that the function $q(x)$ has a horizontal asymptote but this is not at all obvious. To check if, in fact, the function has a horizontal asymptote, you'll change the settings on the x-interval. By widening the x-interval, you'll be able to observe whether the function begins to behave like a horizontal line. Next, you'll widen the x-interval by zooming out in the horizontal direction only.

• Press $\boxed{\diamond}$ $\boxed{\text{WINDOW}}$ and set xScl to 0. Now, press $\boxed{\text{F2}}$, use the down arrow key to highlight SetFactors and then press $\boxed{\text{ENTER}}$. Leave the setting for xFact at 4; change the setting for yFact from 4 to 1. Then press $\boxed{\text{ENTER}}$ twice to return to the graph.

• Press $\boxed{\text{F2}}$ $\boxed{3}$ $\boxed{\text{ENTER}}$ to widen the x-interval by a factor of four. Repeat this process several more times to continue widening the x-interval. Your graph should begin to resemble its horizontal asymptote $y = 2.5$.

CHAPTER 6

Graphing Trigonometric Functions

Three of the six basic trigonometric functions are built-in functions on the TI-89/92Plus/Voyage 200: sine, $\boxed{\text{SIN}}$, cosine $\boxed{\text{COS}}$, and tangent $\boxed{\text{TAN}}$. Before graphing any of these functions, you should first check that your calculator is set to radian mode: press $\boxed{\text{MODE}}$. The Angle setting should be set to RADIAN. (If it is set to DEGREE, highlight that setting, press the right-arrow key and then $\boxed{1}$ for RADIAN, then $\boxed{\text{ENTER}}$.)

Example: Graph $y = \sin(x)$ and $y = \csc(x)$ in the trigonometric viewing window.
- Press $\boxed{\diamond}$ $\boxed{\text{Y=}}$ and erase any stored functions.

- Enter the function $\sin(x)$ as $y1$ as follows:
    On the TI-89, press $\boxed{\text{2nd}}$ $\boxed{\text{SIN}}$ $\boxed{\text{X}}$ $\boxed{)}$ $\boxed{\text{ENTER}}$.
    On the TI-92Plus/Voyage 200, press $\boxed{\text{SIN}}$ $\boxed{\text{X}}$ $\boxed{)}$ $\boxed{\text{ENTER}}$.

Warning! *Since $\csc(x)$ is defined as $\frac{1}{\sin(x)}$, you will raise sin(x) to the -1 power. You could also divide 1 by sin(x). However, you __cannot__ use the* $\boxed{\text{SIN}^{-1}}$ *key which is reserved for the inverse sine function.*

- Enter $\csc(x)$ as $y2$ as follows:
    On the TI-89, press $\boxed{(}$ $\boxed{\text{2nd}}$ $\boxed{\text{SIN}}$ $\boxed{\text{X}}$ $\boxed{)}$ $\boxed{)}$ $\boxed{\wedge}$ $\boxed{(-)}$ $\boxed{1}$ $\boxed{\text{ENTER}}$.
    On the TI-92Plus/Voyage 200, press $\boxed{(}$ $\boxed{\text{SIN}}$ $\boxed{\text{X}}$ $\boxed{)}$ $\boxed{)}$ $\boxed{\wedge}$ $\boxed{(-)}$ $\boxed{1}$ $\boxed{\text{ENTER}}$.

- To graph these functions in the trig viewing window, press $\boxed{\text{F2}}$ $\boxed{7}$ for ZoomTrig. Your graph should be similar to the one in Figure 37.

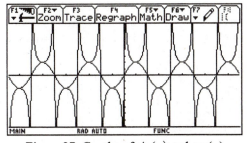

Figure 37. Graphs of $\sin(x)$ and $\csc(x)$

- Press $\boxed{\diamond}$ $\boxed{\text{WINDOW}}$ to observe the settings for the trig viewing window.

The trigonometric viewing window gives a good picture of the graphs of $\sin(x)$ and $\csc(x)$. Keep in mind, however, that it is not the best window for viewing all trigonometric functions. For example, it

137

would not be a good window for the function $y = 5\cos(10x)$. For this function, you would need to adjust the window settings in order to display the key features of its graph.

Approximating Instantaneous Rates of Change

You can approximate the instantaneous rate of change of a function $f(x)$ at $x = a$ by evaluating the average rate of change of the function over a small interval containing a. For example, you might choose a symmetric interval about a of the form $(a - h, a + h)$ for some small value of h. Using this interval, your approximation of the instantaneous rate of change would be

$$\frac{f(a + h) - f(a - h)}{(a + h) - (a - h)} = \frac{f(a + h) - f(a - h)}{2h}.$$

Example: Use the previous formula with $h = .01$ to approximate the instantaneous rate of change of $f(x) = \cos(x)$ at $x = \frac{\pi}{2}$ and $x = \pi$.

On the home screen, you'll enter the formula for the central difference quotient. The key strokes follow. Notice that both numerator and denominator are enclosed in parentheses. The formula is quite lengthy. So, don't press the enter key until you reach the end.

- On the home screen, enter the numerator for the first problem:
  Press (
  On the TI-89, press 2nd COS ; on the TI-92Plus/Voyage 200 press COS .
  Press 2nd π ÷ 2 + . 0 1) .
  Press −
  On the TI-89, press 2nd COS ; on the TI-92Plus/Voyage 200 press COS .
  Press 2nd π ÷ 2 − . 0 1)) .

- Press ÷ .

- Enter the denominator: press (2 × . 0 1)

- Press ENTER . You should get approximately -1.0.

- Press the right or left arrow key. Edit the previous problem by changing $\pi/2$ to π. Then press ENTER . You should get 0. (Can you figure out why?)

Restricting the Domain of a Function

Only one-to-one functions have inverses. In order to define the inverse of a trigonometric function, you must restrict its domain so that the restricted trigonometric function is one-to-one. In the next example, you'll do that for the sine function.

Example: Graph $f(x) = \sin(x)$ on the restricted domain $-\dfrac{\pi}{2} \le x \le \dfrac{\pi}{2}$. Use a $[-2,2] \times [-2,2]$ window.

Note: Your calculator can only interpret one inequality at a time. Hence it cannot interpret the condition $-\frac{\pi}{2} \le x \le \frac{\pi}{2}$ written in this form. Instead, you will need to enter the equivalent condition $-\frac{\pi}{2} \le x$ and $x \le \frac{\pi}{2}$.

- Press $\boxed{\diamond}$ $\boxed{\text{Y=}}$ and clear any previously stored functions.

- Enter the sine function as $y1$: on the TI-89, press $\boxed{\text{2nd}}$ $\boxed{\sin}$ $\boxed{\text{X}}$ $\boxed{\;)\;}$; on the TI-92Plus/Voyage 200, press $\boxed{\sin}$ $\boxed{\text{X}}$ $\boxed{\;)\;}$. Do <u>not</u> press $\boxed{\text{ENTER}}$ because you are not finished defining the function.

- Add the restriction on the domain by pressing $\boxed{\;|\;}$ on the TI-89 and $\boxed{\text{2nd}}$ $\boxed{\;|\;}$ (same key as $\boxed{\text{K}}$ on the TI-92Plus/Voyage 200. Then press the following:
    Press $\boxed{\text{(-)}}$ $\boxed{\text{2nd}}$ $\boxed{\pi}$ $\boxed{\div}$ $\boxed{2}$ to enter $-\frac{\pi}{2}$.
    Press $\boxed{\text{2nd}}$ $\boxed{\text{MATH}}$ (same key as $\boxed{5}$) and then $\boxed{8}$ for Test, $\boxed{4}$ for $\le$, and then $\boxed{\text{X}}$.
    Press $\boxed{\text{2nd}}$ $\boxed{\text{MATH}}$ $\boxed{8}$ $\boxed{8}$ to enter *and*.
    Press $\boxed{\text{X}}$ $\boxed{\text{2nd}}$ $\boxed{\text{MATH}}$ $\boxed{8}$ $\boxed{4}$
    Press $\boxed{\text{2nd}}$ $\boxed{\pi}$ $\boxed{\div}$ $\boxed{2}$ $\boxed{\text{ENTER}}$
    At the end of these key-strokes, your screen should match the one in Figure 38.

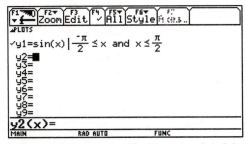

Figure 38. Restricting the Domain of $\sin(x)$

- Press $\boxed{\diamond}$ $\boxed{\text{WINDOW}}$ and adjust the settings for a $[-2,2] \times [-2,2]$ window. Then press $\boxed{\diamond}$ $\boxed{\text{GRAPH}}$.

Notice that $f(x) = \sin(x)$ restricted to $-\frac{\pi}{2} \le x \le \frac{\pi}{2}$ is one-to-one. Hence, it has an inverse.

Do <u>not</u> clear this function from your calculator's memory until after you have completed the next example.

Example: Graph the restricted sine function, $f(x)$, from the previous example and its inverse in the window $[-2,2] \times [-2,2]$.
    The function $f(x)$ from the previous example, should already be stored as $y1$.
    - Press $\boxed{\diamond}$ $\boxed{\text{Y=}}$. If necessary, position the cursor opposite $y2$.

139

- On the TI-89, press ◇ SIN⁻¹ ; on the TI-92Plus/voyage 200 press 2nd SIN⁻¹ . Then press X) ENTER .

- Check that your window settings are for a [-2,2] × [-2,2] window. Then press ◇ GRAPH . Your screen should match Figure 39.

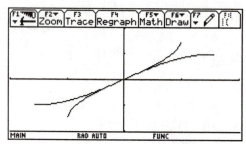

Figure 39. Graphs of $\sin(x)$ and $\sin^{-1}(x)$

Finding Values of Inverse Trigonometric Functions

For the next two examples, check that your calculator's mode setting for Exact/Approximate is AUTO. Press MODE F2 . Change the setting if necessary. Return your calculator to the home screen. Press 2nd QUIT on the TI-89/92Plus and ◇ CALC HOME on the Voyage 200.

Example: Approximate the values of $\cos^{-1}(0.8)$ and $\cos^{-1}(1.8)$.
- To approximate $\cos^{-1}(0.8)$:
  On the TI-89, press ◇ cos⁻¹ . 8) ENTER .
  On the TI-92Plus/Voyage 200, press 2nd cos⁻¹ . 8) ENTER .

  Your answer should be .643501.

- To compute $\cos^{-1}(1.8)$, press the right arrow key to position the cursor at the end of the command line. Press the left arrow key to move the cursor to the left of the decimal point and then press 1 ENTER . This time you'll get an error message. That's because the input value of 1.8 is outside the domain of $\cos^{-1}(x)$. Press ESC .

Example: Find $\tan^{-1}(\frac{1}{2})$ and $\tan^{-1}(\infty)$.
- On the TI-89, press ◇ ; on the TI-92Plus/Voyage 200, press 2nd . Then press TAN⁻¹ 1 ÷ 2) ENTER . The exact answer isn't very helpful in this situation.

- Press ◇ ≈ . You should get .463648. You will refer back to this answer in the next topic.

140

- To enter tan$^{-1}(\infty)$, press the right arrow key to move the cursor to the end of the command line. Press the backspace key, $\boxed{\leftarrow}$, to erase everything to the right of tan^{-1}(. Complete the problem as follows:

On the TI-89, press $\boxed{\diamond}$ $\boxed{\infty}$ (same key as $\boxed{\text{CATALOG}}$) and then $\boxed{)}$ $\boxed{\text{ENTER}}$.

On the TI-92Plus/Voyage 200, press $\boxed{\text{2nd}}$ $\boxed{\infty}$ (same key as $\boxed{J}$)

Then press $\boxed{)}$ $\boxed{\text{ENTER}}$. You should get an exact answer of $\frac{\pi}{2}$.

- Press $\boxed{\diamond}$ $\boxed{\approx}$ for an approximation of $\frac{\pi}{2}$.

Solving Equations Involving Trigonometric Functions

Example: Solve the equation $\cos(x) = 2\sin(x)$.

Recall the syntax of the solve command: solve(equation, variable). You will need to work from the entry line of the home screen.

- Press $\boxed{\text{F2}}$ $\boxed{1}$ for solve.

- Enter the equation:

On the TI-89, press $\boxed{\text{2nd}}$ $\boxed{\text{COS}}$ $\boxed{X}$ $\boxed{)}$ $\boxed{=}$ $\boxed{2}$ $\boxed{\text{2nd}}$ $\boxed{\text{SIN}}$ $\boxed{X}$ $\boxed{)}$. Do not press $\boxed{\text{ENTER}}$ because the command is not yet complete.

On the TI-92Plus/Voyage 200, press $\boxed{\text{COS}}$ $\boxed{X}$ $\boxed{)}$ $\boxed{=}$ $\boxed{2}$ $\boxed{\text{SIN}}$ $\boxed{X}$ $\boxed{)}$. The command is not complete; do not press $\boxed{\text{ENTER}}$.

- Complete the command: press $\boxed{,}$ $\boxed{X}$ $\boxed{)}$ $\boxed{\text{ENTER}}$. On the TI-89/Voyage 200, you get an exact answer; on the TI-92, you get an approximation. (Compare with Figure 40.)

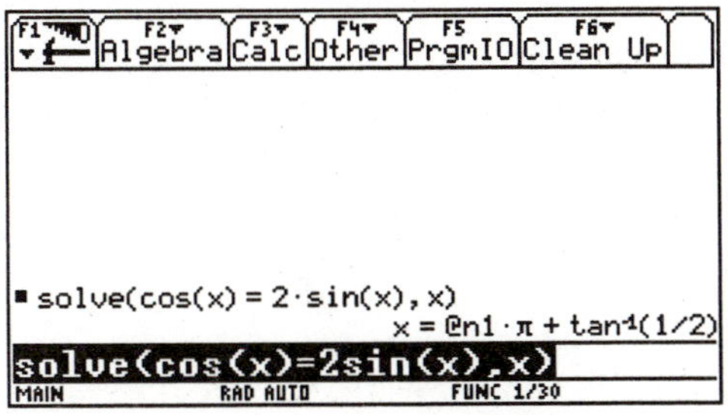

Figure 40. Solving an Equation

Hence, the solution has the form $n\pi + \tan^{-1}(\frac{1}{2})$, where you know, from an earlier example, that $\tan^{-1}(\frac{1}{2}) \approx .464$.

Problems Inherent in the Technology
(Don't Believe Everything That You See!)

Your viewing screen consists of a grid of pixels. When a pixel is *on*, it shows up as a dark square dot on the screen. Graphs are formed by turning on a series of pixels. This method of producing graphs can sometimes produce misleading images.

Example: Graph $g(x) = \sin(x)$ over increasingly wide x-intervals.

- Press ◇ Y= , erase any stored functions and then enter $g(x)$ as $y1$.

- Press F2 7 to graph $g(x)$ in the trigonometric viewing window.

- Turn off the tick marks for the x-axis: press ◇ WINDOW and set xScl = 0.

- Change the zoom factors: press F2 , use the down arrow key to highlight SetFactors, and press ENTER . Set xFact = 10 and yFact = 1 and press ENTER twice to set the factors and return to the graph.

Get ready to have some fun. When $\sin(x)$ is graphed in the trig viewing window, how many complete sine waves do you see. Each time you increase the width of the x-interval by a factor of 10, you should see 10 times as many cycles of the sine wave.

- Press F2 3 ENTER . Do you see 10 times as many cycles? Press F2 3 ENTER again. Do you see 10 times as many cycles as in the previous graph?

The graph produced by the second zoom out shows fewer cycles than the previous graph. In producing the graphs, your calculator does not have enough pixels to capture all the oscillations that are part of the actual graph. In this case, the small subset of points from the actual graph that your calculator chooses to represent with darkened pixels presents a very misleading picture of the features of the actual graph.

CHAPTER 7

Finding Values of Trigonometric Functions

Before solving right-triangle trigonometry problems using the TI-89/92Plus/Voyage 200, you'll need to press ⏢MODE⏢ and change the Angle setting to DEGREE. In addition, check that the setting for Exact/Approximate is AUTO. Begin this section at the home screen.

Example: Compute sin(60°) and sin(29°).
- On the TI-89, press ⏢2nd⏢ ⏢SIN⏢; on the TI-92Plus/Voyage 200, press ⏢SIN⏢. Then press ⏢60⏢ ⏢)⏢ ⏢ENTER⏢. You should get $\frac{\sqrt{3}}{2}$.

- For a decimal approximation, press ⏢◇⏢ ⏢≈⏢.

- Press the right arrow key to move the cursor to the end of the command line. Replace 60 with 29 and press ⏢ENTER⏢. Your calculator will return an exact answer of sin(29). Press ⏢◇⏢ ⏢≈⏢ for a decimal approximation. You should get .48481.

Example: Compute tan(90°).
  On the TI-89, press ⏢2nd⏢ ⏢TAN⏢; on the TI-92Plus/Voyage 200, press ⏢TAN⏢. Then press ⏢9⏢ ⏢0⏢ ⏢)⏢ ⏢ENTER⏢. Your calculator's answer is undef, which means tan(90°) is undefined.

Finding Values of Inverse Trigonometric Functions

Press ⏢MODE⏢ and check that Angle is set to DEGREE and Exact/Approx is set to AUTO before working through the next example.

Example: Approximate the value of sin⁻¹(.3).
  On the TI-89, press ⏢◇⏢ ⏢SIN⁻¹⏢; on the TI-92Plus/Voyage 200, press ⏢2nd⏢ ⏢SIN⁻¹⏢. Then press ⏢.⏢ ⏢3⏢ ⏢)⏢ ⏢ENTER⏢. You should get 17.4576.

Example: Find the approximate value of tan⁻¹(10).
  On the TI-89, press ⏢◇⏢ ⏢TAN⁻¹⏢; on the TI-92Plus/Voyage 200, press ⏢2nd⏢ ⏢TAN⁻¹⏢. Then press ⏢1⏢ ⏢0⏢ ⏢)⏢ ⏢ENTER⏢. Your answer should match the one in Figure 41. Press ⏢◇⏢ ⏢≈⏢ to obtain an approximate solution of 84.2894.

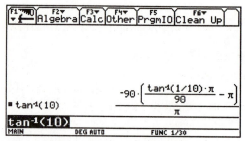

Figure 41. Computing tan⁻¹(10)

Note: *If you are doing a number of problems similar to the last two examples, you may find it easier to change the mode settings for Exact/Approximate to APPROXIMATE. That will save you from repeatedly pressing* ◇ ≈ .

Manipulating Equations Involving Trigonometric Functions

You can use you calculator's computer algebra system to manipulate equations involving trigonometric functions.

Example: Solve $\sin(\theta) = \frac{a}{c}$ for c.

Recall when you use the **solve** command, you must specify the variable for which you are solving. Here's the syntax: solve(equation, variable).

- Press F2 1 for solve.

- On the TI-89, press 2nd SIN ; on the TI-92Plus/Voyage 200, press SIN .

- To create the Greek letter θ: press 2nd CHAR (same key as +) 1 for Greek. Use the down arrow key to highlight θ (it's choice 9) and press ENTER . Then press) = . (You're not done yet.)

- Complete the command:
    On the TI-89, press alpha A ÷ alpha C , alpha C) ENTER .
    On the TI-92Plus/Voyage 200, press A ÷ C , C) ENTER .
  Your screen should be similar to Figure 42.

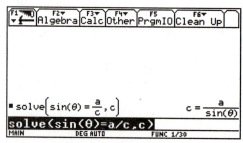

Figure 42. Solving for c.

144

Do <u>not</u> erase the command line until after you complete the next example.

Example: Solve $\sin(\theta) = \frac{a}{c}$ for θ.

Your screen should still match Figure 42.

• Press the right arrow key to move the cursor to the end of the command line. Then change the c that lies to the right of the comma to θ and press $\boxed{\text{ENTER}}$.

• Your screen should be similar to Figure 43. To see the entire solution, press the up arrow key to highlight the solution. Then press the right arrow key to see the rest of this complex solution.

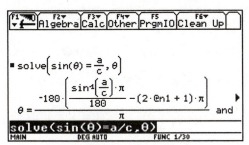

Figure 43. Solving for θ

CHAPTER 8

Working With Parametric Equations

Before you can use your calculator to graph parametric equations, you'll need to change your calculator from function mode to parametric mode. Here's how. Press $\boxed{\text{MODE}}$. The cursor should be blinking on FUNCTION (the present setting for the Graph mode). Press the right arrow key to display the setting choices. Press $\boxed{2}$ for PARAMETRIC. If the setting for Angle is DEGREE, change it to RADIAN. Then press $\boxed{\text{ENTER}}$.

Example: Graph the parametric equations $x(t) = 10 - 3t$ and $y(t) = 1 + 4t$ in the standard viewing window.

- Press $\boxed{\diamond}$ $\boxed{\text{Y=}}$. Notice in parametric mode, you enter pairs of formulas, one for $x(t)$, and the other for $y(t)$. If you have any parametric equations stored in memory, erase them by positioning the cursor on each equation and pressing $\boxed{\text{CLEAR}}$.

- Enter the formula for $x(t)$ as $xt1$: press $\boxed{1}$ $\boxed{0}$ $\boxed{-}$ $\boxed{3}$ $\boxed{\text{T}}$ $\boxed{\text{ENTER}}$.

- Enter the formula for $y(t)$ as $yt1$: press $\boxed{1}$ $\boxed{+}$ $\boxed{4}$ $\boxed{\text{T}}$ $\boxed{\text{ENTER}}$.

- Press $\boxed{\text{F2}}$ $\boxed{6}$ to graph this set of parametric equations in the standard viewing window.

- Press $\boxed{\diamond}$ $\boxed{\text{WINDOW}}$ to observe the settings for the standard window. Your screen should match the one in Figure 44. (If $t$max = 360 instead of approximately 6.28, change your mode settings for Angle to RADIAN.)

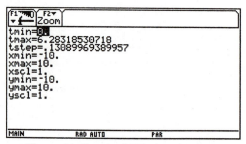

Figure 44. Standard Window Settings

Notice that the settings for the standard viewing window include settings for t: $t$min = 0, $t$max $\approx 2\pi$, and $t$step $\approx \frac{\pi}{24}$. The settings for the x-interval and y-interval are the same as in function mode.

- Press $\boxed{\diamond}$ $\boxed{\text{GRAPH}}$ to return to the graph and then press $\boxed{\text{F3}}$ for Trace. Notice that the trace begins at the point corresponding to $t = t$min, or in this case $t = 0$. Press the right arrow to move along the graph in t-increments of approximately 0.13.

146

Do not clear these parametric functions until you have completed the next example.

Example: Graph the position of a dot as it moves along the path $x(t) = 10 - 3t$ and $y(t) = 1 + 4t$ at one-second intervals from time $t = 0$ seconds to $t = 10$ seconds.

You should already have the parametric equations stored in your calculator from the previous example.

- Press ◇ Y= and highlight the equation for $xt1$. On the TI-89, press 2nd F6 for Style; on the TI-92Plus/Voyage 200, press F6 for Style. Then choose 3 for Square. (If you prefer a smaller mark, choose 2 for Dot instead.)

- Press ◇ WINDOW . Adjust the parameter settings for t: $tmin = 0$, $tmax = 10$, $tstep = 1$.

- Press F2 , use the down arrow key to highlight ZoomFit, and then press ENTER . You should see a group of isolated dots that fall on a line similar to those shown in Figure 45

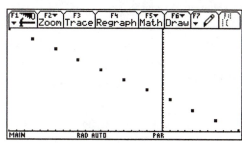

Figure 45. Squares on a Line

- Press F3 for Trace. Then hold down the right arrow key and watch the cursor move from one square to the next.

Using Square Scaling in Parametric Mode

Example: Graph the circle described by the set of parametric equations $x(t) = 5\cos(t)$, $y(t) = 5\sin(t)$, first in the standard viewing window and then in a square-scaling window.

- Press ◇ Y= and clear any stored equations.

- Enter the formula for $x(t)$ as $xt1$ and the formula for $y(t)$ as $yt1$.

- Press F2 6 . In the standard viewing window, your graph looks more elliptical than a circular.

- To change to square scaling, press F2 5 . Your graph should now look like a circle.

147

Combining Two Sets of Parametric Equations

In lab 8B, Bézier Curves, you are asked to form a new set of parametric equations from a combination of two other sets of parametric equations. In the next example, you'll use your calculator to graph such a combination.

Example: Suppose that you have two sets of parametric equations

$$S1:\ \begin{aligned} x_1 &= 2t + 1 \\ y_1 &= -3t + 5 \end{aligned} \qquad S_2:\ \begin{aligned} x_2 &= t - 5 \\ y_2 &= 4t - 3 \end{aligned}$$

and that you want to graph the combination $(1 - t)S_1 + tS_2$ over the interval $0 \le t \le 1$.

You'll tackle this problem in three steps. In Step 1, you'll enter the equations for S_1 and S_2. In Step 2, you'll form the combination. In Step 3, you'll graph the combination.

Step 1: Enter the parametric equations S_1 and S_2.
- Be sure that your calculator is in parametric mode. Press $\boxed{\diamond}$ $\boxed{Y=}$ and clear any stored functions.

- Enter the equations for S_1 as $xt1$ and $yt1$ and the equations for S_2 as $xt2$ and $yt2$.

- Press the up arrow key to highlight $xt1$. On the TI-89, press $\boxed{\text{2nd}}$ $\boxed{\text{F6}}$; on the TI-92Plus/Voyage 200, press $\boxed{\text{F6}}$. Check that the Style is set as Line. (If not, change it to Line.) Do the same for $xt2$.

Step 2: Enter the x- and y-equations for the combination $(1 - t)S_1 + tS_2$:

$$x_3 = (1 - t)x_1 + tx_2$$
$$y_3 = (1 - t)y_1 + ty_2$$

Here's how:
- The cursor should be opposite $xt3$. Enter the equation for x_3:
  Press $\boxed{(}\ \boxed{1}\ \boxed{-}\ \boxed{T}\ \boxed{)}\ \boxed{\times}$.
  Press $\boxed{X}\ \boxed{T}\ \boxed{1}\ \boxed{(}\ \boxed{T}\ \boxed{)}$ for $x_1(t)$.
  Press $\boxed{+}\ \boxed{T}\ \boxed{\times}$.
  Press $\boxed{X}\ \boxed{T}\ \boxed{2}\ \boxed{(}\ \boxed{T}\ \boxed{)}$ for $x_2(t)$.
  Press $\boxed{\text{ENTER}}$.

- The cursor should now be opposite $yt3$. Enter the equation for y_3.
  Press $\boxed{(}\ \boxed{1}\ \boxed{-}\ \boxed{T}\ \boxed{)}\ \boxed{\times}$
  Press $\boxed{Y}\ \boxed{T}\ \boxed{1}\ \boxed{(}\ \boxed{T}\ \boxed{)}$ for $y_1(t)$.
  Press $\boxed{+}\ \boxed{T}\ \boxed{\times}$.
  Press $\boxed{Y}\ \boxed{T}\ \boxed{2}\ \boxed{(}\ \boxed{T}\ \boxed{)}$ for $y_2(t)$.
  Press $\boxed{\text{ENTER}}$.

148